A Handbook of Biological Illustration

Chicago Guides
to Writing, Editing,
and Publishing

(Already Published)

On Writing, Editing, and Publishing
Essays Explicative and Hortatory
Second Edition
Jacques Barzun

Writing for Social Scientists
How to Start and Finish Your Thesis, Book, or Article
Howard S. Becker, with a chapter by Pamela Richards

Chicago Guide to Preparing Electronic Manuscripts
For Authors and Publishers
Prepared by the Staff of the University of Chicago Press

A Manual for Writers of Term Papers, Theses, and Dissertations
Fifth Edition
Kate Turabian, revised and enlarged by Bonnie Birtwistle Honigsblum

Tales of the Field
On Writing Ethnography
John Van Maanen

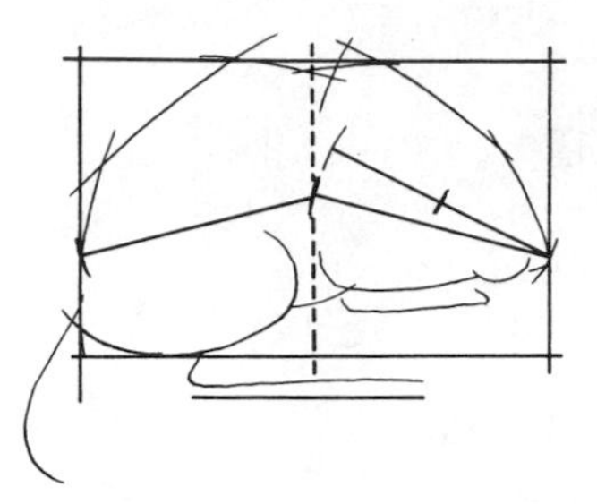

A HANDBOOK OF Biological Illustration

SECOND EDITION

Frances W. Zweifel

THE UNIVERSITY OF CHICAGO PRESS
CHICAGO AND LONDON

FRANCES W. ZWEIFEL is a free-lance biological illustrator whose work has appeared in many journals and magazines. She holds a B.A. in zoology and an M.S. in art and is the author and illustrator of several children's books.

The University of Chicago Press, Chicago 60637
The University of Chicago Press, Ltd., London

 Second edition, published 1988
Printed in the United States of America

97 96 95 94 93 92 91 90 89 88 54321

Library of Congress Cataloging in Publication Data

Zweifel, Frances W.
A handbook of biological illustration / Frances W. Zweifel.—2nd ed.
p. cm.—(Chicago guides to writing, editing, and publishing)
Bibliography: p.
Includes index.
ISBN 0-226-99700-6. ISBN 0-226-99701-4 (pbk.)
1. Biological illustration. I. Title. II. Series.
QH318.Z97 1988
574'.002'1—dc19 87-34271
CIP

IN LOVING MEMORY OF TWO GIFTED TEACHERS

Sister Julitta Larkin,
Trinity College, Washington, D.C.,
AND
Dr. William Homer Brown,
University of Arizona

Contents

Illustrations

Preface

Scholars are trained to analyze words. But primates are visual animals, and the key to concepts and their history often lies in iconography. Scientific illustrations are not frills or summaries; they are foci for modes of thought.

Stephen Jay Gould, "This View of Life: Life's Little Joke," *Natural History* 96 (April 1987): 16

A scientific illustration is the next best thing to holding a specimen in your hand and examining it. Indeed, a drawing may demonstrate relationships more clearly than a trayful of specimens. Illustrations cross language barriers. They not only clarify and augment the text but can reduce the number of words needed.

With these thoughts in mind, the importance of clear, accurate scientific illustrations is obvious. This small book is intended to help the nonartist biologist produce useful—and perhaps even artistically pleasing—illustrations. And I hope that the nonbiologist artist confronted with unfamiliar materials in an unfamiliar field will also find helpful information.

The field of biological illustration has broadened considerably since this handbook first was published in 1961. Since that time old favorite products have disappeared, some techniques are no longer used, and the former most common method of printing for publication has nearly ceased to exist. Other circumstances—more than twenty-five years of new techniques, new printing processes, and an explosion of new products—have urged this long-overdue revision.

For this revised edition I have eliminated the discussion of color techniques, since color rarely is used in scientific journals and the skills required to produce a satisfactory color illustration are beyond the scope of this introductory work. A section on poster presentations has been added, and information on new products and techniques is incorporated into various sections of the manual.

Scope of the Work

BLACK-AND-WHITE DRAWING

The emphasis of this book is on black-and-white drawing, the most commonly used form of biological illustration. By following the suggestions given here, the biological illustrator should be able to proceed from the preliminary sketch to mounting and shipping the finished illustration with a minimum of difficulty. To this end I have stressed time- and labor-saving materials and techniques and small but helpful details. The production of clear and attractive charts, graphs, and maps, so important to scientific papers, is covered along with the more familiar drawing of real objects.

PHOTOGRAPHY

The methods and techniques of photography are not within the scope of this handbook. However, I discuss using photographs as an aid to drawing, retouching photographs, and arranging, cropping, and mounting photographic prints.

Photography and drawing supplement one another as mediums for biological illustration. In some situations the best choice will be a photograph. In other cases a drawing may have several advantages: the lighting and focus may be difficult to achieve in a photograph, giving rise to a false impression; or structures that lack sufficient contrast in a photograph may be shown clearly in a drawing; or inaccurate measurements in a photograph owing to parallax can be detected and compensated for in a drawing. A scientific illustration is not limited to showing what actually exists; when necessary, missing or damaged parts can be reconstructed to show the complete object.

Copying and Copyrights

COPYING

Besides being unethical, copying another's artwork for publication is inartistic. For instance, previous errors may be perpetuated or exaggerated, or new errors may be introduced; the original artist was conveying his or her own impression of a model, which may be very different from the impression the present copyist might have wished to convey from the same model. Also, a copy of any work usually lacks the liveliness of line, freshness, and artistry of the original, as you can demonstrate by copying your own drawings.

COPYRIGHT

Copying directly from another's artwork without permission is illegal. This applies to photographs as well as to drawings and paintings. If it is absolutely necessary to copy from an illustration, as when no model is available to work from, you must obtain permission, in writing and before publication, from the copyright holder (this usually is readily given by scientific publishers). In addition, give credit to the original publication in the legend. If the new illustration is based on—that is, revised from—another's work, note this in the legend by "after [insert original artist's name]." Such crediting is ordinary courtesy and also averts accusations of plagiarism.

A reference collection of photographs and other illustrations to be used for ideas, poses, details of scales, hooves, or feathers, and so on, is a handy and necessary tool for any illustrator. It is not considered copying to refer to others' artwork in this manner.

Acknowledgments

The first edition of this handbook grew out of a thesis I submitted to the Art Department of the University of Arizona. I am grateful to the many persons who helped and encouraged me then, chief among whom were Charles H. Lowe, Jr., Robert M. Quinn, Donald B. Sayner, and the late William H. Brown.

Let me express sincere appreciation to the many persons who have contributed so much help and constructive criticism toward this second edition. Wynne Brown of Brownline Graphics, Knoxville, Tennessee, offered a professional scientific illustrator's wealth of suggestions; Matthew K. Zweifel offered a list of suggestions from the beginning illustrator's point of view and also provided figure 23. I am extremely thankful to Joseph M. Sedaca, graphics manager of the Exhibition and Graphics Department, American Museum of Natural History, for his careful criticism and suggestions. Stephen Jay Gould graciously permitted me to use the quotation at the beginning of this work, which so clearly and succinctly states the reason for doing any scientific illustration. Finally, I am most grateful of all to my husband for his endless editorial tasks and his boundless encouragement—two times around.

1

Printing Processes

It is important that an illustrator understand the printing process the original artwork will undergo. Not every illustration can be reproduced by all processes; some drawing techniques are suited to certain methods of reproduction and not to others. If you know beforehand which process will be used, you can employ the art technique it reproduces best. If cost of publication is an important consideration, it is well to plan an illustration that can be produced by an inexpensive method. It is also an advantage to know beforehand what sort of paper will be used for the reproduction—that is, rough-toothed or glossy—since the surface will affect the reproduction of fine detail. This in turn will influence the amount of time and care devoted to preparing the illustration.

I cannot stress too strongly the importance of knowing beforehand what scientific publication the illustration is intended for. If you are aware of size and technique restrictions before ever setting pencil to paper, you may save hours of work and, possibly, frustration. Some journals publish guidelines to the author and illustrator in each issue, or you can write to the journal editor for specific guidelines.

If it is impossible to find out beforehand the limitations of size and paper surface, then it would be best to work in a standard $8\frac{1}{2}$-by-11-inch format. A simple, strong outline drawing technique without stippling or hatching would suffer the least from reduction in size for publication and would reproduce well on any surface.

Reproduction in Small Quantities

There are several means of reproducing illustrations in small quantities, as for classroom distribution or thesis illustrations. In the latter case, a requirement for permanency will dictate use of a photographic method.

STENCIL COPYING

A familiar name in stencil copying is Mimeograph. This process is suitable for simple illustrations without intricate detail or shading. To copy a printed or drawn work, a stencil must be "cut." Formerly stencils were made on a typewriter adjusted to remove the ribbon from between the type and the stencil. Solid, broken, or dotted lines were drawn on the stencil with special tools. Modern machines "burn" the stencil directly from the copy. All lines and dots on the original artwork must be drawn firmly and sharply in black ink. Shading should be done with stipple and hatching; cross-hatching tends to close up in reproduction.

PHOTOCOPY MACHINES

The office photocopier is a boon to the biological illustrator. Now it is possible to keep a record of an illustration through the stages of its development or in various versions. Some copiers will reduce and enlarge, so you can work up the drawing in a different size from the original sketch without starting all over again. Photocopies are inexpensive, and the better machines will reproduce fine black-and-white detail and even halftones. In fact, there are models that reproduce artwork so well that some scientific journals will now accept photocopies of original illustrations for publication. But check with the publisher first.

PHOTOSTATING

The Photostat, also called a PMT (photomechanical transfer), is a photographic reproduction using a negative made from the original artwork. The process works best with truly black-and-white copy, but halftone work may be screened for printing (see below). A Photostat machine can enlarge or reduce the illustration, but for generally best results a one-third reduction from the original size will clear and reduce imperfections without losing detail.

Illustrations to be photostated may be prepared on any good-quality pure white ground or on tracing vellum or acetate film placed on a white background. (By "pure white" I meant photographically white, which includes light blue but not cream white.) Modern Photostat machines reproduce very sharp copy from strong-contrast illustrations, but gray or weak lines may be blurred or lost. Corrections and parts to be omitted from photostated illustrations should be indicated on the artwork by circling with a blue pencil or by marking corrections on an overlay; they will be painted out of the negative before printing.

PHOTOGRAPHIC PRINTING

If a wash drawing or a photograph is to be reproduced in small quantity, you can photograph it and make a contact print or enlargement from the film negative. Shadow lines from pasted titles may appear in the print as faint gray lines; these can be opaqued on the negative (at extra cost), in which case they will show on the print as pure white lines. All smudges and unwanted pencil lines must be removed from the original before photographing. For photographing line work (black-and-white lines and stipple, for example), special film such as Kodalith will produce sharp contrast without the need for opaquing on the negative.

Reproduction in Large Quantities

Illustrations to be reproduced in large quantities, such as those in scientific journals or books, are printed from plates. There are three classifications of printing plates: intaglio, relief, and planographic.

In intaglio the printing surface is sunken into the plate, and the ink is held in the grooves. Gravure, which produces fine color work but is rarely seen in scientific journals, is an intaglio process.

In a relief process the printing surface is in relief, or raised from the rest of the plate, and the ink is transferred from these raised areas. Rubber stamps and linoleum blocks are small relief plates. Line cuts are printed by metal relief plates. Relief, or letterpress, printing was formerly the most common printing process, but because it takes four times as long as printing by the modern offset process and is therefore more expensive, letterpress has largely been replaced by offset (planographic) reproduction.

In a planographic process the printing surface is the surface of the plate, and water is used to repel an oily ink in the nonprinting areas. Offset printing and lithography are planographic processes. In offset, the metal printing plate is fastened around one of three cylinders of the press. The image to be reproduced is transferred from this metal plate to a second cylinder covered with hard rubber, and the ink is transferred to paper on the third cylinder. Because rubber instead of metal is pressed against the paper in this offset process, a thinner film of ink can be used, yielding a much finer reproduction.

It is possible to print in almost any color ink on nearly any color stock. Scientific journals occasionally publish a map in two or more colors, but in discussing printing processes I will assume black ink on white paper.

For scientific illustration as covered in this book, there are two major divisions of artwork: *line copy,* which comprises only black and white, with no tones of gray, and *continuous-tone copy,* which shades from light to dark through various tones of gray.

LINE COPY

Illustrations containing only absolutely black and pure white areas, no gray tones, may be reproduced either by the relief (letterpress) method of printing or by a planographic (offset) process. Such artwork contains only solid black lines or dots, all shades of gray being represented by stipple or hachure lines or by pattern screens as found in maps and graphs. Illustrations done on special textured grounds, such as coquille board and stipple-surface scratchboards, also may be reproduced as line copy or *line drawings* (see figs. 3, 12, 23, 47, and 54—all examples of line copy).

CONTINUOUS-TONE COPY

In reproduction of continuous-tone black-and-white illustrations, which contain tones of gray from light to dark, the print itself still consists of black ink on white paper. The gradual shading effect is achieved by the *halftone screen.*

Photographs and continuous-tone drawings (pencil, wash, carbon dust) are photographed through a fine screen, which breaks up the light into tiny dots. The dots will be largest in dark areas of the drawing, smaller in gray areas, and microscopic in the lightest areas. If you look closely, you may be able to see these dots in a newspaper print. The greater the number of dots per inch, the more delicate the gradations of shading. In most scientific publications, fine screens are used (150 to 200 lines to the inch). Screens up to 300 are possible, but they require high-quality paper to do them justice, increasing the cost of reproduction (see fig. 1 for examples of halftone screens).

Because the screen covers the whole negative, no absolute white is reproduced by a halftone plate. If pure white is required, the white areas must be dropped out by the printer (*dropout halftone*). Likewise, silhouetting the figure to give a pure white background (*outline halftone;* see fig. 60) requires extra work by the printer. These operations add to the cost of printing. Gradation that is barely discernible in the original may not reproduce perfectly in the halftone, since dot formations tend to weaken delicate work. Very fine lines will appear as broken lines or may even be lost in reproduction. Lightly shaded pencil drawings are especially difficult to reproduce well—some printers will not accept them. In general, continuous-tone drawings will reproduce better by offset than by letterpress.

An attempt to make a halftone reproduction from a previously published halftone illustration is usually unsuccessful, owing to the moiré effect caused by screening a screen, which may range in appearance from "watered silk" to an obvious checkerboard (see fig. 2).

A *combination plate* is a page of illustration that includes both line and continuous-tone copy. Such a plate must be photographed twice, once for each kind of copy. This will increase the cost of printing, but sometimes no other method of reproduction will do, as when the halftone figure or the line work, or both, must have a white background. If a faint tone in the background is acceptable, however, the cost may be kept down by reproducing the whole plate as a halftone, lettering and drawn lines as well as the continuous-tone drawing or photograph. To do this the illustrator must place all lettering inside the bounds of the halftone copy (see fig. 59A). Because everything in the illustration will be photographed through a screen, everything will be printed in tiny dots of ink, including the continuous-tone work, numbers, lettering, symbols, and lines.

Shadow lines from the edges of pasted-on lettering will show up in a halftone. The combination plate, though more expensive than the halftone plate, thus looks more professional (see fig. 59B). The labels and numbers are placed on an overlay and photographed separately as a line shot; halftone and line cut are printed together (see p. 96 for overlay directions).

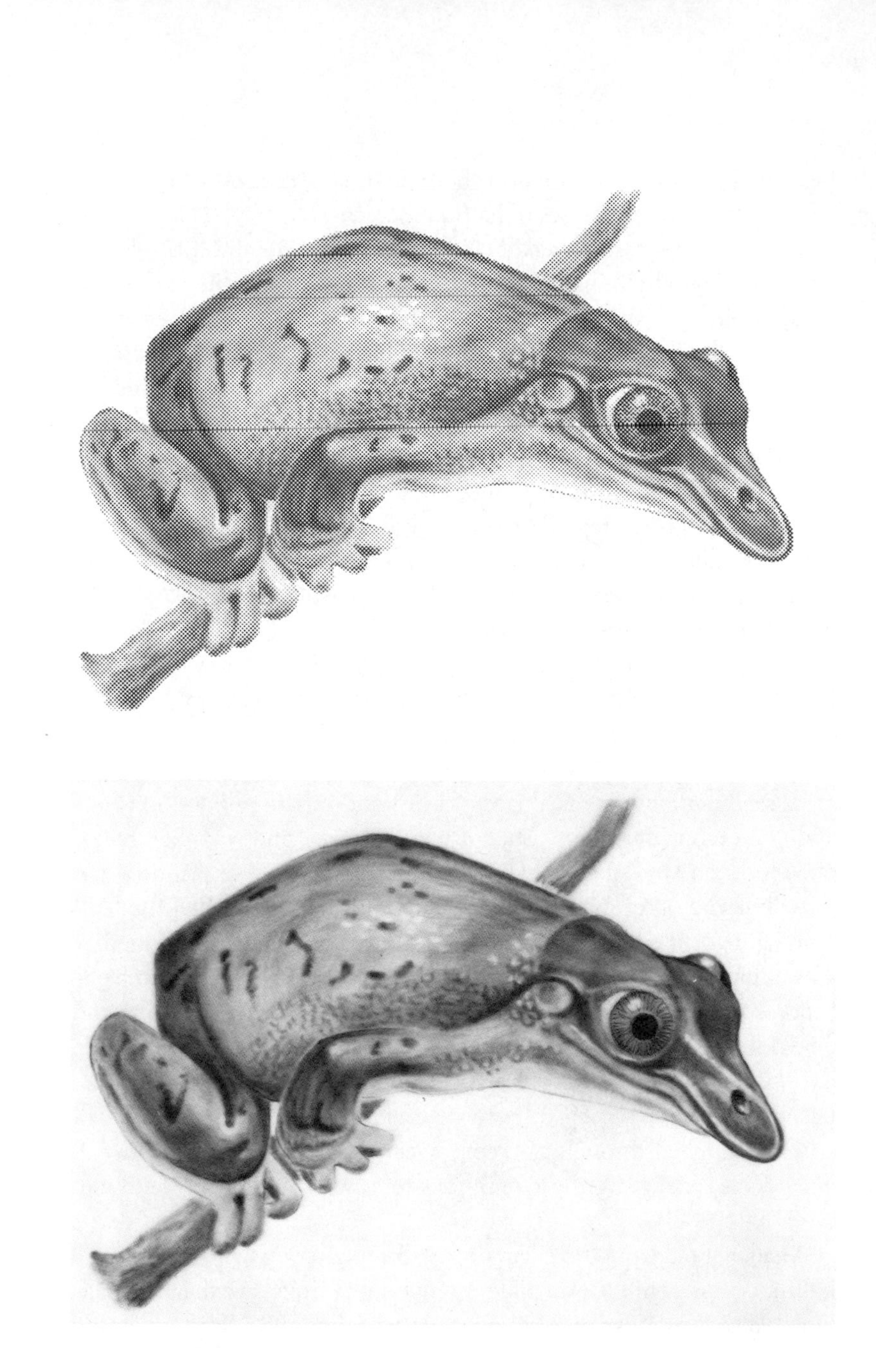

FIG. 1 A halftone illustration, showing the effect of two different screens, 65 and 133 lines per inch. Reduced slightly from original carbon-dust drawing of the frog *Triprion spatulatus*.

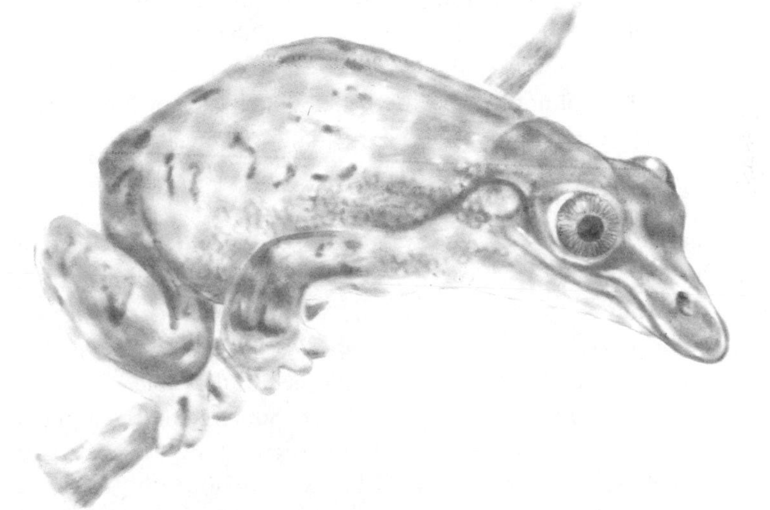

FIG. 2 A, Enlargement of small section of figure 1, showing a magnified view of a halftone screen; example taken from the eye of previous figure. B, figure 1 printed to show moiré effect resulting when a halftone illustration is made from halftone copy.

2

Size and Reduction of Illustrations for Publication

Advisability of Reduction

Ordinarily, black-and-white illustrations, such as line drawings, maps, and graphs, are reproduced smaller than the original copy (the artist's drawing), for sound reasons. When working in a larger size, the artist can draw with more detail and precision. Also, slight reduction of the original will conceal small imperfections of line and increase the sharpness of the illustration in general. This does not mean you can rely on reduction to make your work satisfactory; true, a slight reduction may improve a good picture, but nothing short of burning will help a poor one.

Photographs—especially those with intricate detail—may be reproduced in the original size, though some publishers prefer to reduce them slightly. Continuous-tone illustrations such as pencil and wash drawings may benefit from a slight reduction for publication.

Planning the Copy Size

The amount of reduction the illustration must undergo will depend on how large the artist likes to work, on how much detail and fine drawing are included in the picture, and on what the editor of the publication decides. In general a reduction of *one-quarter off* (75% of the original size) to *one-third off* (66⅓% of the original size) is desirable, because it cleans the line work without loss of detail. Some editors may prefer more or less reduction, but they seldom if ever want more than half off.

For a reduction of one-quarter off draw your original illustration exactly one and a third times as large as you want it to appear in print. To prepare an illustration for a one-third reduction, draw one and a half times as large as the desired final printed copy. For a one-half reduction the original should be twice as large as the print will be.

When giving reduction directions, it is most important to understand what you are asking. "Reduce one-half" means the same as "one-half off" or "reduce to one-half"; the final size will be one-half the dimensions of the original drawing, or one-fourth the original area. "Reduce one-third" and "one-third off" do *not* mean the same as "reduce to one-third." A drawing measuring 3 by 6 inches would measure 2 by 4 inches

reduce to 4 inches

FIG. 3 An acceptable method of indicating amount of reduction desired. Reduced slightly from the coquille board and wax crayon drawing of the caterpillar *Nematocampa limbata*.

if the reduction directions read "reduce one-third" or "one-third off"; the same drawing would measure 1 by 2 inches if the directions read "reduce to one-third." The wording of the reduction directions therefore must be carefully checked.

The desired reduction usually is written in the margin of the artwork in blue pencil. If the illustration will not be reduced to a convenient fraction of its original size, the directions may read "Reduce to *x* inches" (or millimeters) along a line drawn in the appropriate margin. (Only one edge need be marked, since the proportions remain constant in reduction.) Some editors prefer that the reduction be stated in terms of printer's linear measure, points and picas, rather than in inches or millimeters. Six picas equal 1 inch, and 12 points equal 1 pica. A reduction direction properly given is shown in figure 3.

Before ever beginning to work on an illustration, the artist should examine a copy of the journal in which it will be published. Measure page size and study the format to anticipate some of the size limits placed on the published illustration. In publications with a two-column format, cost considerations may require that figures be reduced to the width of a single column or else fit neatly across two columns. Ignorance of the space allowance may render an otherwise adequate illustration virtually useless through the loss of detail by too great a reduction (see fig. 4). If you are aware beforehand of the size of the published result, you will be able to tailor the size and detail of your original to the restrictions imposed by the medium of publication.

A professional editor can be expected to require that illustrations be adequate for the method of reproduction employed. Unfortunately, the editors of most biological journals are not professionals and usually have that unremunerated task added to numerous other academic and scientific duties. They often lack both the time and the ability to properly evaluate the graphic material that passes through their hands. It then becomes the artist's responsibility to see that the artwork is suitable in all respects for the method of reproduction and size of illustration to be used. The wise illustrator therefore will examine a copy of the intended journal's guidelines before planning the first pencil stroke.

In determining the size of an illustration that will occupy a full printed page, remember to allow for the legend. In an imaginary case, the allowable plate measurements are 6 by 8 inches. Two lines of legend will occupy one-quarter of an inch (the amount of space per line of legend will, of course, vary with the type size used in the journal), and you should allow at least $\frac{1}{8}$ inch of space between the legend and the bottom of the illustration. Subtract this total $\frac{3}{8}$ inch from the vertical inches available for the plate, leaving $7\frac{5}{8}$ inches (high) by 6 inches (wide) for the illustration. Since a one-half reduction is desired in this hypothetical case, the working size will be twice the final measurements, or $15\frac{1}{4}$ by 12 inches.

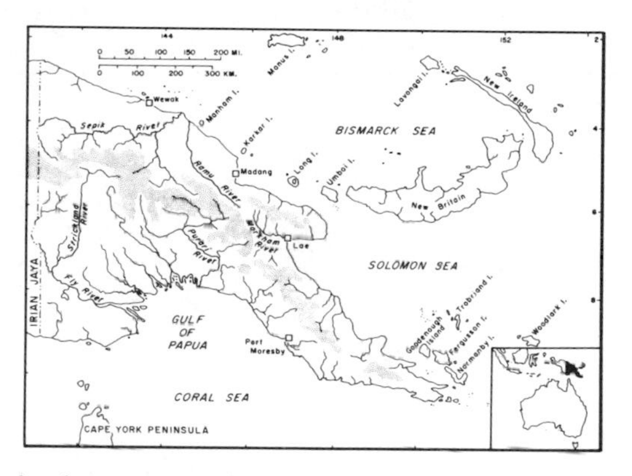

FIG. 4 Example of too great a reduction, from an original approximately 10 inches wide to the customary $2\frac{1}{2}$ inch single-column width, with consequent loss of detail and legibility.

One way of planning the working size is to use a *copyfit rectangle*, as shown below (fig. 5). First, draw a rectangle the size of the reduced illustration (rectangle *ABCD*), extending side *AD* to twice its length, making line *DAE* (this will be the only measurement required). Extend side *DC* also. Draw a diagonal line from *D* to *B* and extend it beyond *B*. At point *E*, draw a horizontal line to intercept the diagonal (point *F*), and from this point drop a perpendicular line to the extended line *DC*; the dropped perpendicular will intercept the extension of *DC* at point G. The proportions of the new rectangle (*EFGD*) are the same as the proportions of the original rectangle, but the new rectangle is twice the size of the original.

The copyfit rectangle is a convenient method of checking the proportions of the finished original against the desired reduction size. Place a tracing of the smaller rectangle (*ABCD*) in one corner of the finished illustration (*EFGD*). The proportions are correct if a rule laid diagonally across the illustration also lies diagonally across the reduction-size rectangle.

An inexpensive and very worthwhile investment is a plastic circular "slide rule" calibrated specifically for determining reduction size and proportions. This instrument reads directly in inches, number of times

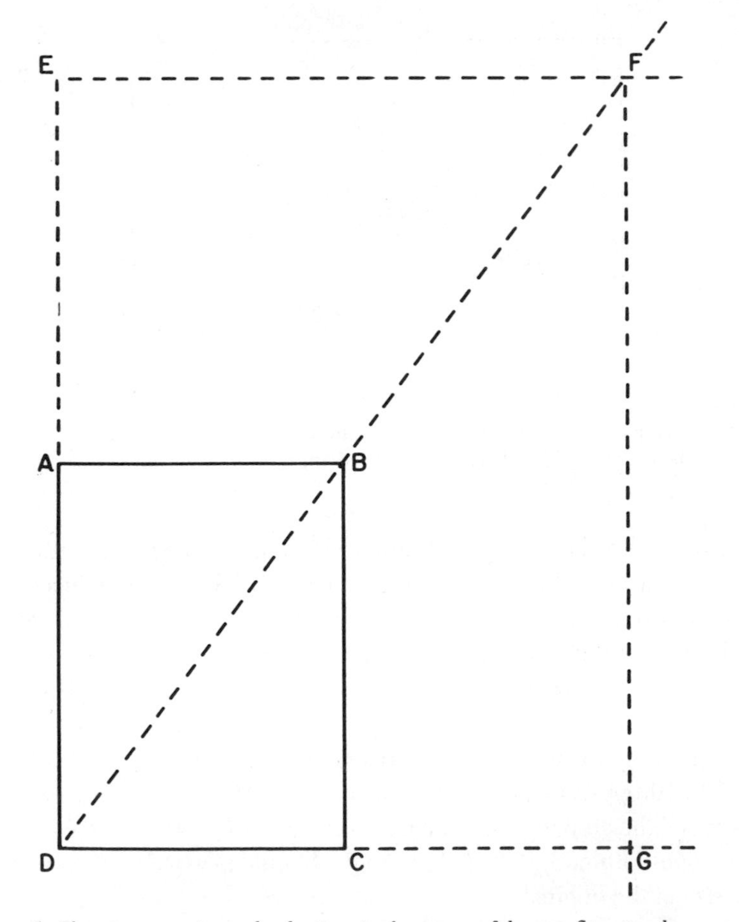

FIG. 5 Planning copy size and reduction size by means of the copyfit rectangle.

of reduction, and percentage of original size. In a typical use, for example, you can set the rule for the desired reduction (percentage of original size) and then read the correct size of the original directly opposite the size of the reduction desired.

To save cost, all the illustrations for one paper or article should be made to undergo the same amount of reduction. If a single uniform reduction is not possible, a minimum number should be planned (each variation in reduction size requires resetting the camera, increasing the cost of reproduction). In addition to the financial consideration, it is artistically pleasing to have continuity in lettering size throughout the published work.

If the figures on a particular plate must undergo different reductions, the illustrator will have to make a dummy plate. Draw the true plate dimensions on a white board; then draw rectangles representing the final, reduced figures in place. These rectangles must be numbered to correspond to separate numbers on the individual figures. The numbers, titles, and so on, are written in place on the dummy. Reduction directions, plate and figure numbers, author's name, and title of the paper or book should be written on the back or in the margin of each illustration, and all are then put in an envelope attached to the dummy plate.

A reducing lens is invaluable when working on any illustration that is to be reduced in publication. By viewing the illustration through this lens from time to time while working, you can better judge how heavy to draw the lines, how much shadow to use in certain areas, how much detail may be included, and so forth. But though the reducing lens will give some idea of the final reduced illustration, keep in mind that the printing process tends to make lines somewhat thicker and more closely spaced than the view through the lens suggests. For this reason, draw in a slightly more open manner than you actually desire for the printed result—place lines and dots farther apart and draw shadows not quite so dark in the original as you may want them to appear.

An alternative method of determining whether an illustration will reduce satisfactorily is to reduce it on an office copying machine. If the stipple and hachure hold up well in a one-third or one-half reduction by this method, there should be no problem in reduction for publication.

3

Materials

The illustrator should be properly equipped before beginning to draw. Basic and special materials are listed below, under the techniques in which they are used. Note that several items appear under more than one heading. Some items are optional; their purchase depends on individual need and preference.

It would be to your great advantage to visit a large art-supply store before making any special purchase. New products come on the market almost daily; alas, tried-and-true old favorites disappear just as quickly. Spend a few hours examining what is on the shelves and asking questions. Often the clerks are art students themselves and can suggest new and convenient products. If no good art-supply store is within reach, many will mail their catalogues; a short list is included in the Selected References at the end of this book.

An important consideration is that many art products are hazardous to health. Fixative sprays may be dangerous to inhale; dyes and even drawing surfaces may give off poisonous fumes; inks can cause some users skin and eye irritations. Be sure to follow all cautions printed on product labels, and be prudent about spraying any product in a closed area. Many adhesives are highly flammable, and some of the newer grounds will not only go up in a flash of flame but give off toxic fumes while burning. Modern illustrators must be prepared to protect themselves from the latest wonders of the art world. To this end, you are strongly advised to read one or both of the books by Michael McCann listed in Selected References.

Many of the newer, more convenient products mentioned below are intended for commercial art and are not designed to last indefinitely. Some adhesives and films turn yellow within a year, some become brittle and eventually crumble away, others dry and curl up. If permanency is desired, buy the highest-quality paper, preferably pure (or mostly) rag, and ask for archival-quality mounting boards and cover papers. The only adhesives that will last through the years without yellowing, staining, or drying out are glues made of organic products, either vegetable or animal. (For a discussion of such products see Ralph Mayer, *The Artist's Handbook of Materials and Techniques*, in Selected References.) Most scientific illustration, however, is intended for reproduction and therefore need not last more than a year or two. For these, many of the convenient (if short-lived) products will do the job, and with added speed and ease.

Preliminary Drawings (Rough Drafts)

- Lead pencils ranging from hard to soft (4H to 4B)
- Artgum, Pink Pearl, or white vinyl eraser
- Kneaded (charcoal) eraser
- Ground: layout paper (any inexpensive ground, such as newsprint, bond, computer printout discards; tracing paper is most convenient)
- Pencil sharpener or sandpaper tablet

Ink Drawings

- Several straight penholders
- Pen points of various sizes and degrees of flexibility, such as fine mapping, crow-quill, and writing points
- Black waterproof drawing ink for the pens above
- Technical pens with various points
- Special inks for technical pens, such as Pelikan brand
- Chamois or lintless cloth for wiping pen points
- Grounds: opaque papers such as plate-finish Strathmore, hot-press bristol board, or illustration board; tracing vellum such as Albanene, Bienfang's Satin Design, or Denril drafting paper/film; transparent matte-surface acetate or polyester film
- Erasers: kneaded (for cleaning the ground); Koh-I-Noor "imbibed" eraser (for vellum and film)

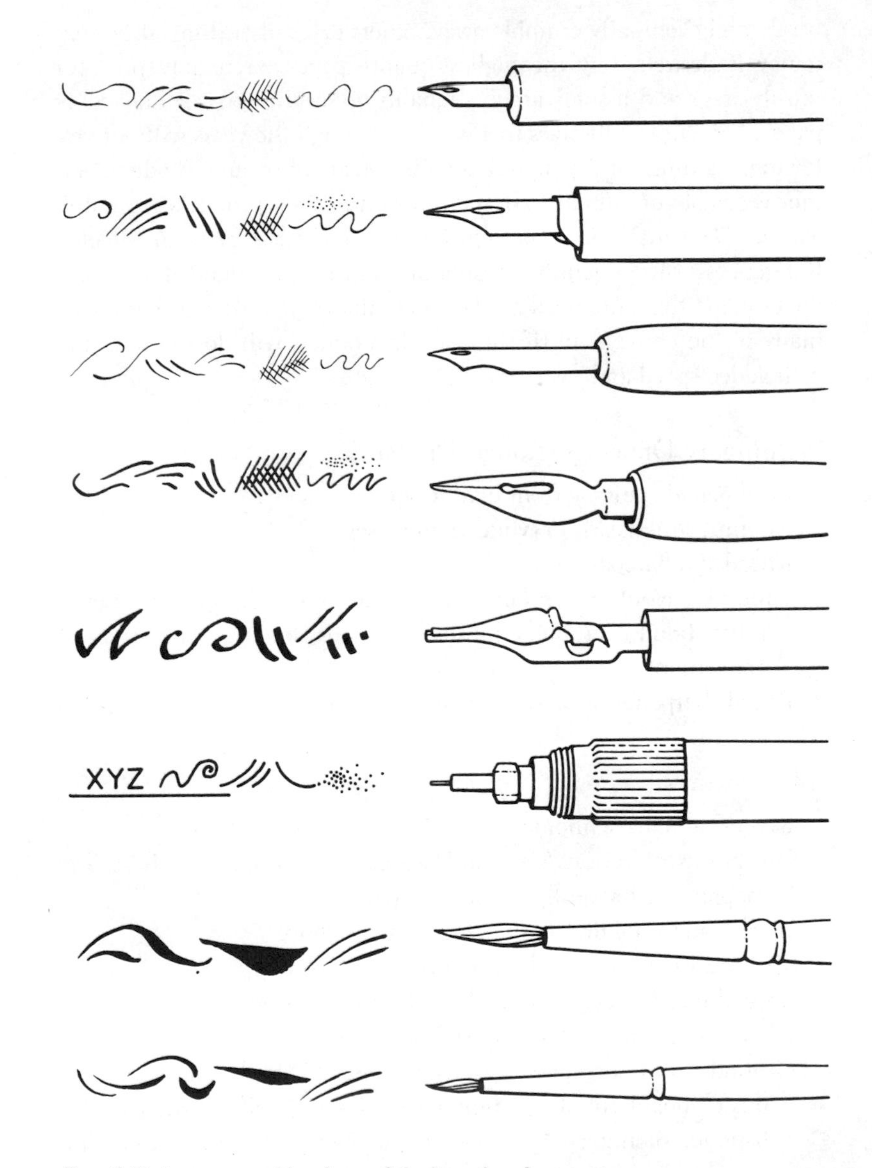

FIG. 6 Various pens and brushes and the lines they draw.

Fine scalpel or X-Acto knife (blade no. 11 or 16) for cleaning lines on film

Gouache white-out paint such as Pelikan Graphic White, Dr. Martin's Bleed Proof White, or Pro White

Scratchboard Drawings

Fine and medium round watercolor brushes

Medium-fine and medium-coarse pens

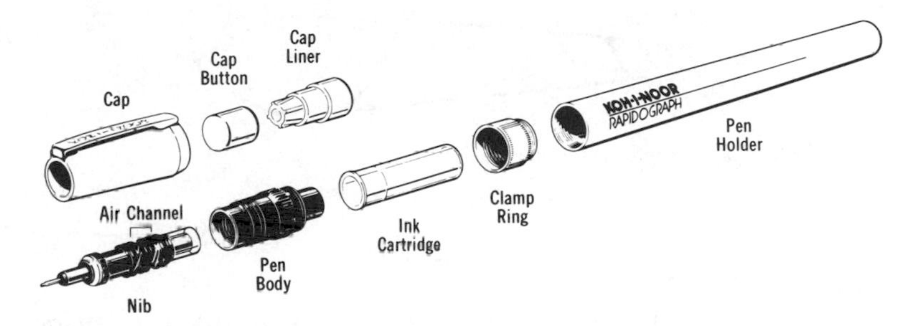

FIG. 7 Technical pen. (Courtesy Koh-I-Noor Company)

Waterproof black drawing ink; special inks for technical pens

Scratchboard, such as EssDee British scraperboard (source given in Selected References); *or*

Matte-surface (frosted) acetate or polyester film, heavyweight

Erasers: kneaded (for cleaning the ground); Koh-I-Noor "imbibed" eraser (for film)

Pounce (draftsman's pumice)

Fine scalpel or X-Acto knife (blade no. 11 or 16)

Stiff mounting board

Adhesive tape or white artist's tape for mounting

Special Black-and-White Techniques

Coquille board

Stipple-surface EssDee British scraperboard (source given in Selected References)

- Black wax lithograph crayons, soft, medium, and hard; or black "all surface" pencil
- Medium-fine and medium-coarse pens
- Waterproof black drawing ink
- Gouache white-out paint such as Pelikan Graphic White, Dr. Martin's Bleed Proof White, or Pro White
- Fine scalpel or X-Acto knive (blade no. 11 or 16)

FIG. 8 Rolled-paper smudger (tortillon).

Pencil Drawings

- Lead pencils ranging from hard to soft (4H to 4B), wooden or mechanical
- Kneaded eraser
- Grounds: cold-press (toothed, rough-surface) papers, bristol board, illustration board; hot-press (smooth surface) papers; drafting vellum such as Bienfang's Satin Design, Albanene, or Denril; matte-surface acetate or polyester film
- Rolled-paper smudgers (tortillons)
- Erasers: kneaded, white vinyl

Wash Drawings

- Round and flat watercolor brushes
- Permanent black watercolor in a tube, such as Winsor & Newton
- White china saucer or palette
- Containers for water
- Absorbent tissues or towels
- Ground: illustration board
- Drawing pen and ink
- Kneaded eraser

Carbon-Dust Drawings

Carbon (charcoal) pencils, such as Wolff's, Conte, or General's, in soft, medium, and hard (B, HB, H)
Round watercolor brushes, large and small
Flat watercolor brushes, various sizes
Small covered dish

FIG. 9 French curves.

Sandpaper pad
Chamois, pieces of chamois
Rolled-paper smudgers (tortillons)
Kneaded erasers
White vinyl eraser
Ground: vellum-surface bristol paper; best-quality scratchboard; matte-surface (frosted) acetate or polyester film
White backing for transparent ground
Penknife to sharpen carbon pencils, vinyl eraser
Fixative spray

Maps and Graphs

Grid paper ruled with blue lines
Ground: tracing vellum such as Bienfang's Satin Design, Albanene, or Denril; or matte-surface polyester film

Technical pens of various sizes
Ink for technical pens
Kneaded eraser for cleaning the ground
Koh-I-Noor "imbibed" eraser
Rules: inch and/or metric scales
French curves or ship's curves; flexible curved rule
T-square or triangles
Stencil with various-sized circles, squares, and triangles

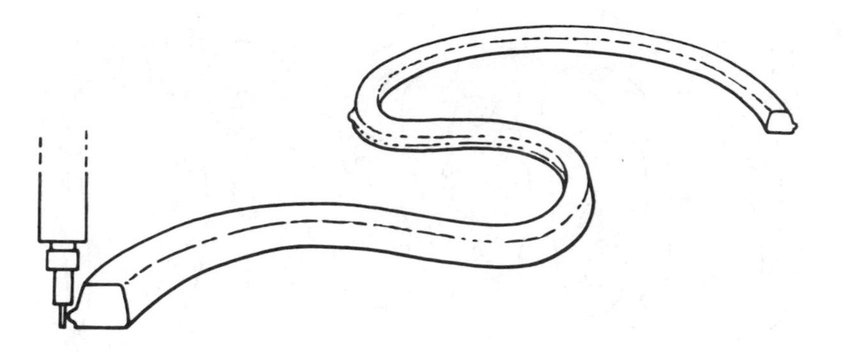

FIG. 10 Flexible curved rule.

Shading (pattern) film such as Letratone, Chartpak, or Formatt
Fine scalpel or X-Acto knife (blade no. 11 or 16)
Gouache white-out paint such as Pelikan Graphic White, Dr. Martin's Bleed Proof White, or Pro White

Lettering

Lettering device such as Leroy lettering system; *or*
Plastic lettering guides such as Wrico; *or*
Press-on or dry-transfer lettering such as Letraset, Chartpak, or Formatt
Technical pens and ink
T-square
Nonskid (cork-backed) metal rule, 18 or 24 inches long
Lead pencils, hard and medium

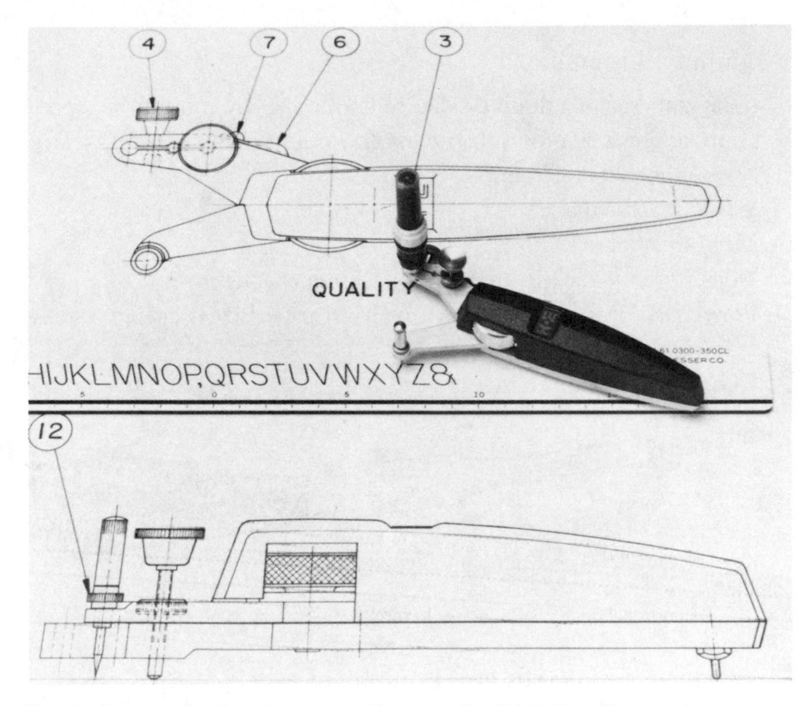

FIG. 11 Leroy scriber lettering system. (Courtesy Keuffel & Esser Company)

Kneaded eraser
White artist's tape
Scotch 811 Magic Plus removable tape
Pounce (draftsman's pumice)

Retouching Photographs

Retouch paints, such as those made by Grumbacher and Pelikan, *or*
Black and white gouache paints
White china dish for mixing paints
Round watercolor brushes
Lead pencils, medium and soft (HB, B, 2B)
White Conte pencil
Charcoal pencils
Sandpaper pad

Mounting Illustrations

Spray adhesive; *or* double-sided self-adhesive dry mount; *or* repositional pressure-mount tissue; *or* dry-mounting tissue with electric iron; *or* rubber cement
White artist's tape
Mounting board
Nonskid (cork-backed) metal rule, 18 inches long
Protective cover such as brown or bond paper, clear acetate sheets, tracing vellum
Rubber-cement thinner

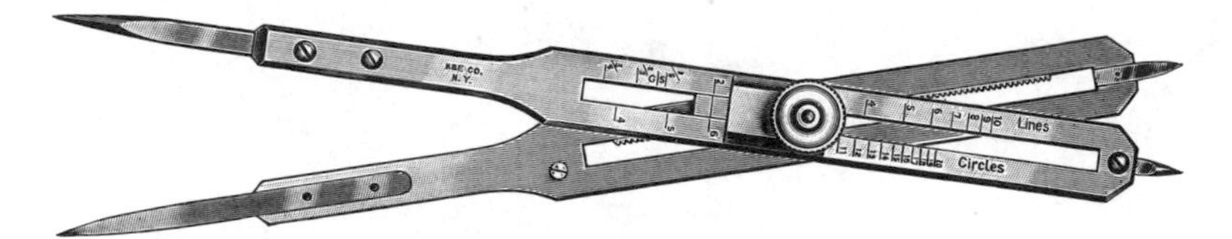

FIG. 12 Proportional dividers. (Courtesy Keuffel & Esser Company)

Miscellaneous Materials and Equipment

Below are some special items that may be useful, depending on their availability, the amount of illustrating to be done, and the financial resources of the artist.

Binocular magnifiers (on a headband)
Calipers
Metric scale for measuring specimens
Small forceps
Simple dividers
Proportional dividers
Telescope
Grid-scored Lucite for telescope
Ocular reticule and/or micrometer for microscope
Opaque projector
Camera lucida
Microscope camera lucida
Mechanical typesetters (Kroy, Merlin, etc.) or computer typesetter

4

The Preliminary Drawings

Introductory Remarks

The purpose of a biological illustration is to clarify or demonstrate what is contained in the written work. To accomplish this end, there are several guiding principles that are applicable to all biological illustrations executed in any style or medium.

OBSERVATION

The most important quality in a scientific artist is well-developed powers of observation. Faulty observation, which leads to faulty illustration, usually is caused either by a lack of understanding of the subject or by "too much" understanding. Happily, the first is often corrected by the attempt to draw the object, which leads to further study and knowledge.

The second source of trouble—too much understanding—is not so easily recognized or corrected. In this instance one may tend to represent what one knows about the object or believes to be true rather than what actually exists. What is needed in this instance is a new perspective, which may be gained by so simple a means as examining the subject in an unfamiliar position. Try looking at it upside down or reflected in a mirror.

CLARITY

The illustration must be neat and complete but not crowded. Including more detail than necessary is unwise and wasteful. The purpose the illustration is to serve, or the persons who are to use it, will determine how detailed the work should be. A type specimen, for instance, might be drawn in full detail, whereas the major differences among several genera may be better illustrated by a series of simple line drawings.

Do not try to include too much information in the illustration; this may only confuse the viewer. A graph with a great many lines may be useless, and a drawing so detailed that it distracts attention from or disguises the important point is worse than useless. An illustration prepared for a high-school textbook might differ in detail from one intended for a scientific journal.

EMPHASIS AND COMPARISON

The illustrator often is inclined to exaggerate a point under discussion, but this may give a false impression and should be avoided. It is possible to give prominence to particular features by the use of certain techniques and thus to capture attention without distortion. For instance, in drawing comparisons between two or more subjects, emphasize differences not through exaggeration but through accentuation, by darkening lines in critical areas and by using shading and highlights that throw these areas into relief.

SHADING CONVENTION

In scientific illustration, a drawing customarily is shaded as if the light source (or principal light source) is just out of the upper left corner of the picture. This convention serves three purposes: this is a common position of the actual light source when one is comparing a specimen with an illustration; there appears to be a common light source for several illustrations that may be grouped together on one plate; and it is easier to compare drawings by different artists if the light source appears the same in all drawings. Occasionally special lighting is required to emphasize some particular detail, in which case the illustrator should recognize the shading convention for what it is—a convention, not a law.

In general, double lighting of an object is desirable because it makes the picture more interesting in value and form and keeps it lighter than would a single source of illumination—a dark drawing does not reproduce as well as one lighter in tone. The second light source is not as strong as the principal one and may be placed wherever the artist chooses.

PERSPECTIVE

"Aerial perspective" means the artistic expression of space by loss of definition, value range, and color warmth owing to intervening atmosphere between the observer and the object. The greater the distance, the more obvious these effects will become. Imagine moving away from nearby mountains. At close range you can see canyons and crags, a pattern of highlights and shadows, and various warm colors. As you move away, the mountains seem to lose their ruggedness, the value pattern becomes one continuous tone, and the colors become cooler. From a great distance the mountains look flat and textureless, with no range in value, and the color is uniform violet or blue.

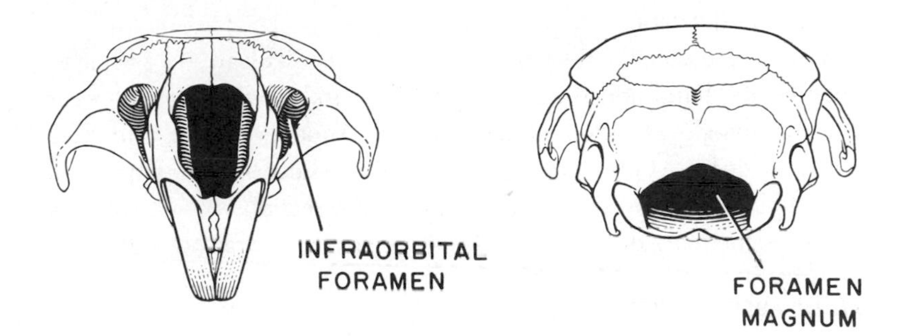

FIG. 13 Aerial perspective in a simple line drawing. Heaviest lines are used in the closest parts, and the drawn lines are broken where a more distant part of the skull passes behind a closer part. Note the treatment of the arrows within the drawing. Reduced one-third from the original scratchboard drawing of the giant rat, *Xenuromys barbatus*.

Showing aerial perspective in black-and-white drawings is not at all difficult. The nearer parts are given the deepest shadows and lightest lights, the most detail, and the heaviest outlines. Even simple line drawings will exhibit a three-dimensional effect if the heaviest lines are used to draw the closest parts. Another means of showing perspective is to break the line of the distant part that passes under or behind a closer part just at this junction (see fig. 13). The greater the distance between the two parts, the greater should be the break in the line.

Objects in space appear to shrink as they recede, parallel lines seem to converge, and horizontal lines do not remain horizontal. Drawings, to be convincing to the viewer, must at least approximate these conditions. To achieve this effect, the artist applies linear perspective. Ordinarily the biological illustrator will not be required to produce an elaborate picture such as a landscape, requiring the use of linear perspective. If a knowledge of linear perspective is necessary, refer to a textbook or see John Montague, *Basic Perspective Drawing: A Visual Approach*, listed in Selected References.

CLEANLINESS

Keep the work as clean as possible, since dirt and smudges may reproduce photographically. If such marks cannot be erased satisfactorily from line copy, they may be covered with superwhite gouache paint (see p. 67, Methods of Correcting Mistakes), or with strips of white paper pasted over the area.

One way to keep the work free of dirt is to place a clean sheet of paper between the hand and the drawing while working. Some artists wear white cotton gloves with the fingers cut out. Another method is to use a mask: cover the whole drawing surface with a sheet of plain paper and cut a hole in this top sheet just the size of the section you are working on. Tape the mask down so it will not slide over the drawing surface.

ALLOWABLE LATITUDE

The latitude allowable in scientific illustration may permit some deviation from absolute precision except in significant details. For example, you might draw the outline of a skull in a free manner, while a certain feature of that skull that is being discussed must be rendered in absolute detail. The purpose the illustration is to serve also may determine the latitude permitted. A picture used as an embellishment or to show an animal in general will not be drawn as precisely as an illustration of a particular structure in a particular specimen. Common sense will tell you how much freedom you may exercise.

ARTISTRY

An illustrator will find it an advantage to study drawings by other artists. You must determine how much detail to include and how much you can leave out for the most effective illustration. Be cautioned against that most common pitfall, overworking the drawing. Enough is good; a little more may ruin the picture.

Every effort should be made to produce a picture that is artistically pleasing as well as biologically correct. Not everyone will achieve technique of professional quality, but there are many things you can do to produce an attractive illustration. Neatness counts: make all lines clean and clear, and be sure there are no smudges, fingerprints, or blots on your drawing. Try to achieve lifelike poses of live subjects and at least graceful positions of dead ones. When grouping drawings together on a plate, maintain balance by placing heavier, darker figures at the bottom and lighter figures at the top, if possible. Fine-lined drawings usually do not group attractively with heavily drawn figures. The amount and percentage of blank areas on a plate—the "negative space"—also may be important to the overall appearance.

Often a plain outline drawing may be made more interesting merely by adding a background line to suggest the horizon; this will give depth to the picture. The use of shading film around the simply drawn subject may seem to lift the subject from the paper surface. Such devices may be used tastefully to liven a simple line drawing (see fig. 20).

Making the Preliminary Drawing

The first rough ideas for the illustration are sketched on any inexpensive layout paper. Typewriter bond, blank sheets from the computer printer, even newsprint may be used.

A most convenient sketching ground is tracing paper, which is relatively inexpensive, has a smooth surface, and takes erasure well. The chief advantage in using tracing paper for sketching is that you can draw and retrace a drawing, changing and rearranging lines and details, over and over again until you have produced a satisfactory preliminary sketch. The tracing paper may be taped over a printed grid (such as graph paper), to make measuring easier. Some artists do all their sketch corrections on one piece of tracing paper by flipping the sketch over and altering the lines on the back, then flipping back again, erasing the front and altering again, until the sketch is satisfactory. This procedure requires working in mirror image half the time. Alternatively, you may want to keep a record of your sketch progression and therefore may choose to use separate numbered tracings.

Measuring the Subject, Using Common and Special Drawing Aids

Large objects may be defined as those that can be seen clearly and in sufficient detail with the unaided eye. Small objects are those that, though not microscopic, are so small that some magnification is necessary to show detail. Microscopic objects require considerable magnification (as with a compound microscope) to reveal detail.

MEASUREMENTS AND PROPORTIONS

Before beginning to draw, the illustrator will have determined the dimensions of the drawing to allow for the reduction in publication (see chap. 2). Unless you employ a special technique, such as the use of the telescope (see below), you will have to measure your subject and adjust the measurements to suit the desired reduction or enlargement of the drawing.

Convenient measuring instruments, in addition to the familiar inch or millimeter rule, are calipers, simple dividers, and proportional dividers. Commonly, calipers are available with a maximum linear measuring distance of 6–8 inches, though larger tools are made. Measurements may be made directly from calipers, whereas a measurement laid off with dividers must be read by resting the tips of the dividers against a scale. In using either of these instruments, you determine the final reduction or enlargement in the illustration by multiplying the measurement of the object by the reduction or enlargement factor. This may be done arithmetically, or you may employ a graphic method such as that described next.

On a piece of graph paper draw a horizontal line (fig. 14, *AB*). Erect a perpendicular at *B* equal in length to the largest measured dimension of the object (fig. 14, *BC*). Erect another perpendicular (*BD*) that bears the proper multiple of *BC* to agree with the desired proportional change in the drawing (in this case, $\frac{2}{3}x$). Now, in order to convert any measurement of the object to the proper corresponding dimension in the drawing, merely lay that measurement off on the graph perpendicular to *AB*, as at 1–2, and read the new dimension opposite, as at 2–3.

The use of proportional dividers eliminates the need for arithmetic or graphic calculation of dimensions for the final drawing. Proportional dividers (fig. 12) consist of two legs, pointed at both ends, on an adjustable sliding pivot. The dividers may be set so that the distance between the points at one end bears the desired relation to the distance between the points at the other end. With the pivot locked in place, the proportions remain the same with the dividers open or closed. Thus, if the instrument is set to multiply a linear measurement two and one-half times, a measurement of 6 millimeters taken at the smaller end will measure 15 millimeters at the larger.

The illustrator should "block in" the outline of the object being drawn. For instance, if you wish to draw twice natural size, draw a box twice the length and twice the width of the actual object. Within the box, mark off various reference points, such as half the object length,

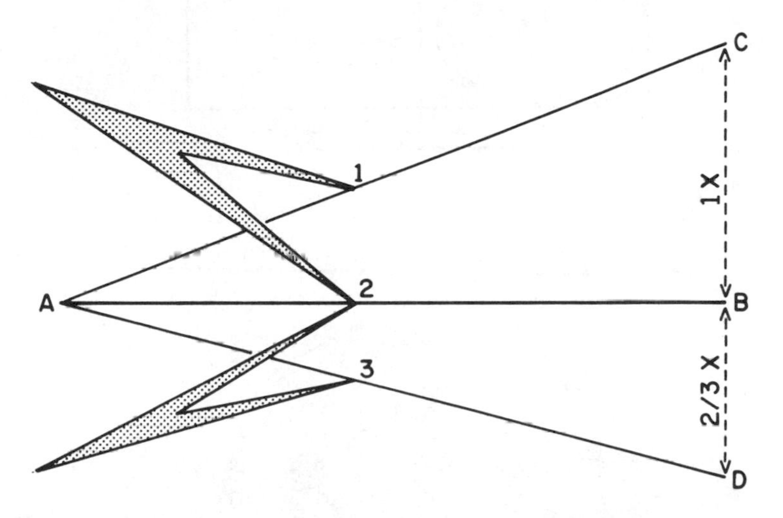

FIG. 14 Enlarging or reducing measurements by using simple dividers.

one-quarter of the object length, length from posterior margin to a particular feature, and other similar points (fig. 15). Do not add together separately measured parts of the subject to compute the total length, for the accumulation of small errors in measurement may make a large error in the overall length. It is better to take all measurements starting from one end of the subject.

In drawing bilaterally symmetrical objects, you can save time and effort by drawing only one half. The drawn half is then traced and transferred to the mirror-image position of the drawn half. After transferring the sketch, measure carefully to be sure that the transferred drawing is accurately placed.

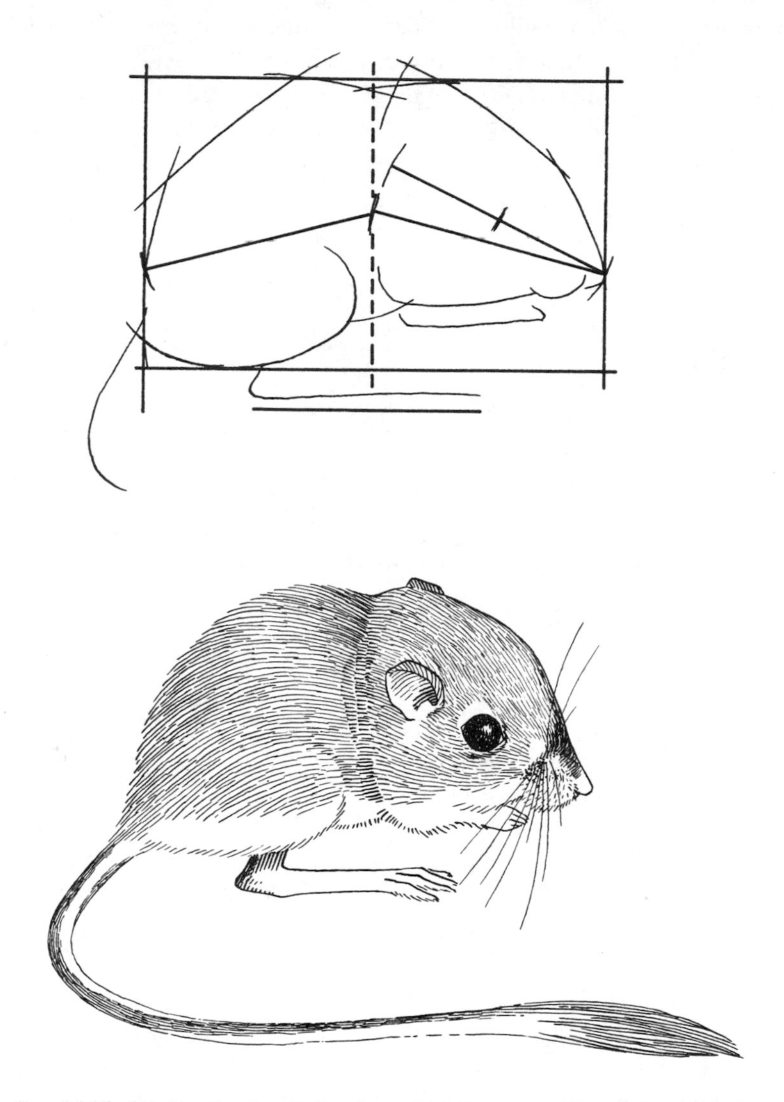

FIG. 15 Blocking in a drawing. Reduced one-third from original pen drawing of the kangaroo rat, *Dipodomys merriami*.

You may encounter several difficulties when attempting to measure a three-dimensional object for a two-dimensional drawing. (Biologists unfamiliar with problems of illustrating sometimes assume they can take exact measurements from the drawing of the object, without considering the foreshortening that is bound to occur in making the drawing.) All measurements should be taken in the same plane in order to illustrate *what is seen*. In figure 16, line *AC* represents the picture plane, or plane of measurement for the top, and line *AD* is the plane for

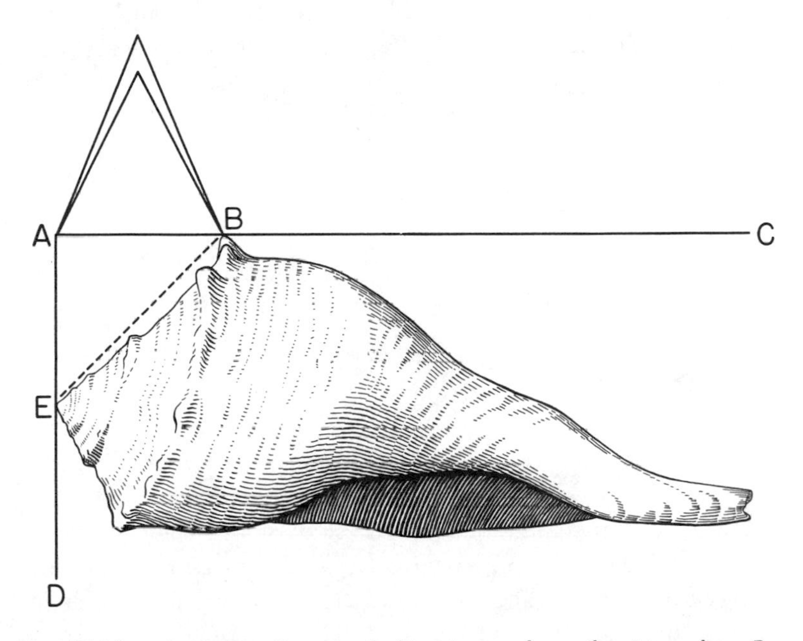

FIG. 16 Measuring a three-dimensional object in one plane—the picture plane. Reduced one-half from original pen drawing of the lightning whelk, *Busycon sinistrum*.

the end view of the shell. It is obvious that measurements not in the picture plane do not remain proportional in the drawing because of curvatures of the shell and relative distances from the plane of measurement. Therefore, in the finished rendering of the shell's top view, actual distance *EB* would be represented as distance *AB*.

Binocular vision causes a stereoscopic effect because each eye has a slightly different angle of view. Closing one eye may provide a better perspective (monocular) as you take measurements and do the preliminary sketch. After blocking in the total length and width and indicating other necessary points of reference, move back and forth in order to position your eye directly over each section as you draw it. If you keep your eye stationary, foreshortening may cause difficulty in measuring (see fig. 17A). The farther the object is from your eye, the less serious will be the problem; but then the object may be difficult to see clearly.

By viewing the specimen through a telescope, you will reduce the foreshortening, see the object more clearly, and at the same time achieve a favorable monocular perspective (fig. 17B). In addition, by placing a

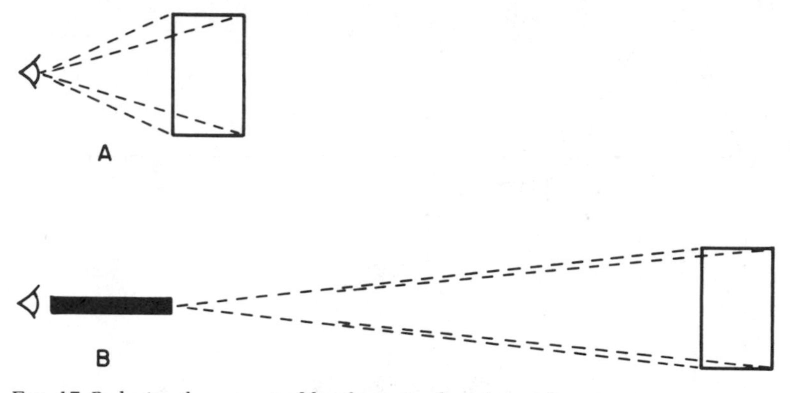

FIG. 17 Reducing the amount of foreshortening by using a telescope.

grid in front of or behind the object to be illustrated and drawing on suitably marked graph paper, you can avoid much measuring and computing of proportions. The telescope need not be expensive; a child's simple instrument will do.

Mount the telescope firmly on a stand so that the eyepiece is at a convenient height. The object to be drawn is placed against a piece of graph paper held vertical to the long axis of the telescope or directly behind a sheet of clear Lucite scored in a grid. You can then view the object through the telescope and draw on a piece of graph paper.

The object grid coordinates (on graph paper or scored Lucite) and the drawing paper grid coordinates are keyed with letter and number series in vertical and horizontal directions so that a point on the object grid can be duplicated with a corresponding point on the drawing paper

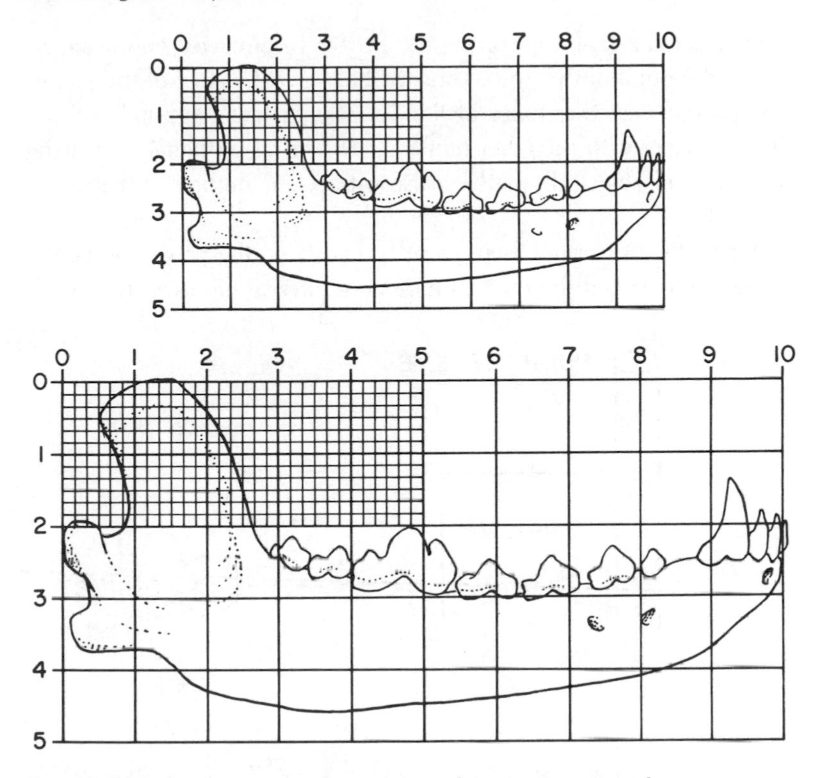

FIG. 18 Using coordinate squares to increase or decrease proportionately.

grid (see fig. 18). Using the appropriate coordinates, the features of the subject as seen through the telescope can be transferred to the drawing paper with a minimum of perspective distortion. The degree of reduction or enlargement of the object is determined by the ratio of object grid size to drawing paper grid size.

The telescope itself keeps the eye from wandering from a stationary position and also allows the artist to see detail clearly. The use of this simple procedure has proved invaluable in saving time and achieving accuracy.

THE CAMERA LUCIDA

The camera lucida consists in essence of a four-sided glass prism mounted on a stand that holds it above the drawing paper. The image of the object to be drawn appears to the artist to be projected on the paper. When the pupil of the eye is directed half over the edge of the prism, the observor sees the image of the object with half the pupil and the drawing paper with the other half (fig. 19). This simultaneous viewing of paper and object allows the artist to trace the outline of the image with fidelity.

Since the eye cannot focus simultaneously on the paper and pencil at one distance and the object image at a different distance, the object

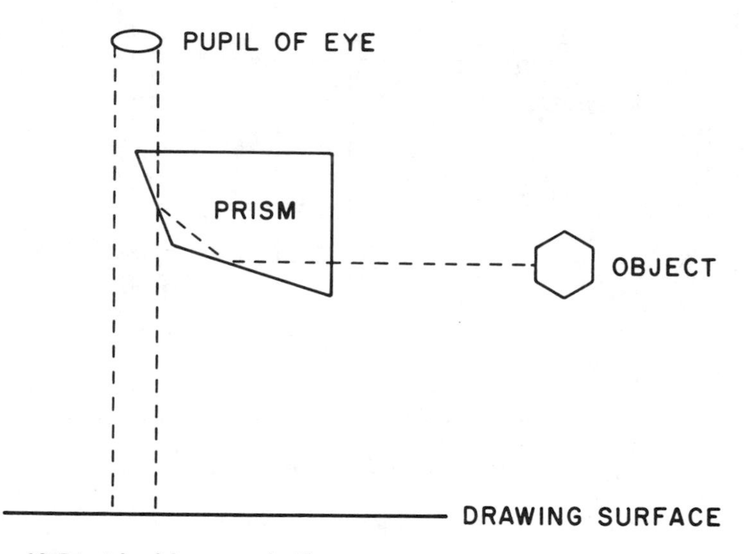

FIG. 19 Principle of the camera lucida.

must be placed the same distance from the prism as the prism is from the drawing paper. Thus the drawing will be the same size as the object. To produce a drawing at a different ratio, the object must be moved closer to or farther from the prism; a plus or minus lens is placed in front of the prism to correct the focus. If the image is not properly focused on the paper, the pencil point will appear to "crawl," leading to eyestrain and errors. To avoid distortion, be sure to align the paper, prism, and object properly. Check the alignment by mounting a small cube or straight-sided object in the position of the object to be drawn. If the image shows only the top of the test cube, alignment is proper. If the image is off center so that you can see a side of the test cube, shift it until only the top shows. The possibility of distortion in a camera lucida drawing is high; it takes special care to produce a drawing of proper proportions.

Camera lucidas may be purchased from large art-supply stores or perhaps may be borrowed from a university art department. To use a microscope camera lucida, see page 36.

PHOTOGRAPHY

An accurate outline for a subject may be obtained by photographing the subject, enlarging or reducing the photograph, and then tracing the photograph.

MICROSCOPES

A dissecting microscope (also called a "stereoscopic microscope") will be useful for seeing and drawing small specimens. You can sketch the outline by means of the telescope and fill in the details under the microscope; or if the specimen is small enough to fit entirely within the microscope field, you can draw both outline and details while viewing the object through the microscope. View the object under the binocular microscope with only one eye for drawing in order to see it without the problem of perspective. It may be necessary to prop the object slightly on an angle so the plane of view will be at a right angle to the object plane for monocular viewing.

A compound microscope is used to see and draw objects not discernible with the naked eye. Use a scanning lens to locate the portion of the slide to be illustrated, then use lenses of greater magnification to see the desired part in greater detail. Filters of different colors will let you see certain particulars more clearly by bringing out one stain while subduing others.

An ocular micrometer or, preferably, a grid-ruled reticle that fits into the ocular of either a binocular or a compound microscope will enable you to draw accurately by using the grid-coordinate method as described on page 33.

Microscopes with continuously variable magnification ("zoom scopes") are especially useful when drawing with an ocular grid, since the size of the image can be adjusted precisely to the dimensions of the grid. Then draw the subject by the grid-coordinate method.

To determine the magnification of the object drawn under the dissecting microscope, divide the total length of the object into the total length of the drawing. Another way to determine the magnification of a microscopic object is to use a precalibrated ocular micrometer; compare the measurement made with the ocular against the corresponding measurement of the drawing.

In figuring the magnification of the drawn object, keep in mind the reduction the artwork will undergo in publication. For this reason it may be best (and easiest) to indicate the size of the object as follows. Draw the object by the coordinate-squares method (see fig. 18), using a grid-ruled ocular in the microscope and graph paper for the drawing surface. When you have completed the work, keeping the same magnification you used while drawing, place a millimeter rule in the microscope field of view. Measure how many squares of the ocular grid fit into a convenient measurement, such as 5 millimeters, on the rule. Transfer this scale to your drawing. Now, no matter how the drawing later is reduced or magnified, the proportions will remain the same, and there will be no need to juggle measurement calculations.

MICROSCOPE CAMERA LUCIDA

The microscope camera lucida is an expensive item, but most university biology departments have one. The device is attached above the eyepiece of a microscope and, by means of a broken-beam prism and a mirror, lets you view the specimen through the microscope and at the same time see an image of the specimen apparently projected on a sheet of paper. The image is traced on the paper as you look into the microscope. By varying the objective and the eyepiece magnifications, you can make the image larger or smaller. The cautions about properly balanced lighting and alignment of elements made concerning the camera lucida (p. 33) also apply here. On older models that lack a lock for the mirror position, check the angle of the mirror by placing a small cube or cylinder (end up) on the drawing paper and manipulating the mirror until you see only the top, not the sides, of the test object.

When measuring the specimen under the microscope and transferring this measurement to the drawing, be sure to hold the scale (millimeter rule) at the same level as the specimen, *not flat on the microscope stage*, or an error will result.

SUPPORTING THE OBJECT FOR DRAWING

Most objects must be held in the desired position while being drawn. One convenient method is to mold a support of children's oil clay or Plasticine. Soften and shape the clay by squeezing it between your fingers and press the specimen into position on the clay. A neutral color is best, since colors may cast undesirable reflections or stain the specimen. Heat softens the clay and may make it cling to the object, so heat sources such as lamps should not be placed too close.

Another convenient holder for small specimens is a dish of clean sand. Use white or black sand to create a contrasting background. Such a support can also be used under fluids if necessary. A tiny specimen can be fixed to a large marble with wax, oil clay, or soluble glue and the marble can be rotated in the bed of sand to the desired position.

In an emergency, paper towels may be substituted for oil clay or other support. Soak the towels, then crumple them into a more or less solid mass and allow them to dry in the desired position. To support very small objects, a chewed-up bit of paper towel or even a kneaded eraser may suffice.

Drawing Live Animals

Live models present many difficulties, the chief being motion. As in any drawing from nature, you must spend at least as much time observing the subject as you spend drawing it. Sitting quietly near the animal for a half-hour or so will enable you to observe its form and manner of movement; also, it will give the animal a chance to become accustomed to your perhaps unfamiliar presence. Some animals grow nervous, even hostile, when they are stared at and may be less than cooperative. You may have to look away occasionally or leave the area for a while. Also, smiling so as to show one's teeth is not advised; an animal subject may interpret this as threatening behavior.

It may be necessary to anesthetize an animal to keep it quiet enough to sketch, though you must be careful not to introduce an unnatural posture. Amphibians and reptiles may be chilled into quiescence by placing them in a refrigerator for a short time; carbon dioxide will anesthetize insects without permanent harm if not applied for too long; a small amount of chloroform will make some animals drowsy, and merely feeding others will calm them. Birds may be kept in the dark until they are needed for actual drawing; usually the sudden light will keep them quiet for a while.

Special pens and cages may be built or improvised for immobilizing live animals. The material of the cage must not hide the animal; one side made of glass or Lucite may be sufficient. Small aquariums and wide-mouthed glass jars covered with hardware cloth can be used as cages. A pane of glass can be placed inside the aquarium to confine the subject to a smaller area. Generally speaking, the animal should have enough room to turn around and assume a comfortable position without being either cramped or allowed to roam.

Precise details may be studied while the animal is more closely confined. Some lizards and snakes can be held in one hand while you draw them with the other, but small mammals are likely to bite unless held in a manner that prevents close scrutiny. The human hand is too warm to hold an amphibian in comfort for any length of time. Wide-mouthed quart and gallon jars and small aquariums will confine such animals for precise observation and drawing of details.

Subdued or colored lights may be useful when studying and sketching nocturnal animals.

To capture the peculiar characteristics of the animal to be illustrated, make many sketches from all possible angles. Begin by blocking in the simple form and add detail and shading later (fig. 15). Blocking in and measuring the animal by head lengths or foot lengths are useful shortcuts. The head, feet, and other particulars should be practiced from different angles. Only after a great deal of quick sketching will an illustrator be familiar enough with the subject to make a creditable drawing.

The Polaroid camera can be an invaluable aid in depicting live animals. Snapshots of the subject will augment the artist's memory of color patterns, ear position, and so on. Still shots of the animal in motion can help in producing lifelike scientific illustrations.

Fur resembles masses with soft outlines. Occasional hairs may be suggested, but do not become so involved in representing individual hairs that you lose the form and character. Grounds with a textured surface, such as coquille board and EssDee patterned scraperboard, are useful for depicting fur. The scraperboards may be scratched with a sharp point to suggest hairs.

Only the outlines of feathers are seen from a short distance, and even these outlines disappear as the bird recedes from the observer. Scales may be suggested in medium-value areas and not depicted at all where there are highlights and shadows unless the scale pattern is the important part of the illustration.

Enlarging or Reducing the Sketch

You may find that the finished sketch is not the optimum size to work with. Perhaps you discover it will have to be reduced more than one-half for publication; or it may be too small to contain all the necessary detail. There are several ways to change the size of the sketch without redrawing it from the beginning.

Modern photocopy machines, which not only copy clearly but also enlarge and reduce, are a blessing to the scientific illustrator. A photocopy of the sketch also can be enlarged or reduced as desired.

By using the coordinate-squares method, you can enlarge or reduce a simple drawing (see fig. 18).

An opaque projector, used to show diagrams and art examples during lectures, can be useful to project (and enlarge) a sketch for retracing. A slide projector can project an image on a wall; the distance from the wall determines the size of the image. When using any of these devices, carefully position the angle of projection to avoid distorting the proportions of the original drawing.

A large-model camera lucida or opaque projector, sometimes known as a Goodkin or Laci-Luci, may be found in some university art departments. Such a machine projects an image of two-dimensional copy onto a drawing surface and, by means of chains and wheels, can enlarge or reduce this image as desired.

When complete, the preliminary drawing should contain all necessary details and desired shading. The next step is to transfer this finished sketch to the ground chosen for the final rendering, as discussed in chapter 5.

5

The Finished Drawing

Once you have made a satisfactory preliminary sketch, you must transfer it to the ground to be used for the finished drawing. The choice of technique will dictate the choice of ground, which will in turn determine the method of transfer.

Transferring the Drawing

If the final ground is transparent, tape it to the preliminary sketch so it will not slip out of position and trace the sketch lightly with a medium to hard lead pencil (H to 4H). Some transparent grounds, such as acetate and polyester films, may pick up the sketch in mirror image on the back. To prevent this, you may need to spray the preliminary drawing with fixative before you begin to trace it; or make a photocopy of the sketch and trace that.

If the sketch must be transferred to an opaque ground, you can choose from several transfer methods. One of the best is the double-transfer method, which produces a copy exactly like the original. The original drawing must be done with a pencil soft and dark enough to allow two transfers—B or HB usually will do. Place the original copy face down on a sheet of tracing tissue and rub firmly all over the back of the drawing with a smooth burnisher, transferring the graphite from the drawing to the tracing paper (the image will show on the tracing paper in mirror image). Now repeat this process, transferring the tracing-paper image to the final ground.

In the single-transfer method, blacken the back of the original sketch with a soft graphite pencil (3B), tape the sketch right side up on the final ground, and retrace the original lines with a sharp-pointed, hard lead. (If you use a colored lead for this, you will be able to see just which lines you have already transferred.) If the preliminary sketch was done on tracing paper, there is no need to blacken the entire back—just turn the sketch over and blacken the drawn lines on the reverse side. A difficulty with either of these single-transfer methods is that the tracing rarely has the fidelity or freshness of the original sketch. On the other hand, whereas single transfer is messier than double transfer, it has the advantage that it can be used to transfer a sketch done in hard pencil or in ink, crayon, or even felt-tip pen.

To save the original sketch, trace it and use this tracing as directed above. Or place a thin sheet of graphite transfer paper between the rough sketch and the final ground. You can make your own by blacking a sheet of thin paper with a soft lead or use purchased transfer sheets such as Saral brand. Do not use typewriter carbon; this is almost impossible to erase.

A tracing table, or light box, will let you copy the sketch onto most drawing papers, such as bristol or two-ply plate-finish paper for ink renderings. The tracing table consists of a sheet of frosted glass placed over a shallow box containing bright lights. If this is unavailable, you can improvise: tape the sketch and final ground to a window on a bright day.

Some grounds, such as soft papers, scratchboard, and coquille board, are delicate; use light pressure when transferring the sketch to such a surface. Always use as light a touch as possible when transferring for a pencil drawing, since any indented line on the final ground will interfere with future carbon deposit and may show up white on the finished drawing.

The final ground should be absolutely clean. Before transferring the sketch to it, you may need to erase the new surface completely with an Artgum ("soap") eraser or a kneaded ("charcoal") eraser, depending on the fragility of the ground. Remove all eraser crumbs, then transfer. After transferring the sketch, erase all smudges, fingerprints, and unwanted pencil marks with a kneaded eraser or with a special "imbibed" eraser for acetate and polyester film, as needed. (The Koh-I-Noor Rapidograph "imbibed" eraser is impregnated with a chemical that removes ink from synthetic grounds without damaging the surface.) Protect the final ground from further smudging and skin oil by placing a piece of paper under your hand while drawing, or make a mask for the drawing (see p. 26).

When the preliminary drawing has been transferred satisfactorily to the final ground, you are ready to begin the finished drawing.

Black-and-White Illustrations

Line copy is any artwork that contains only black and white, no gray or color; the effect of shading is achieved by the placing and spacing of lines and dots. Line copy is usually reproduced by offset, sometimes by letterpress.

PEN-AND-INK DRAWINGS

Use good quality for the final ground. Hot-press (smooth) bristol board, plate-finish drawing paper, and illustration board are all satisfactory. There are newer products in use, such as flexible acetate and polyester sheets, frosted on one side to receive pencil, crayon, or pen. Good-quality vellum tracing paper is also recommended. It is advisable to find out what method of printing your illustration will undergo, if possible, since some methods will require your using one of these flexible grounds.

You should have several straight penholders and a variety of crow-quill pen points, a bottle of waterproof black drawing ink, and a piece of chamois or other lint-free cloth for wiping the pen points. Technical pens also may be used for drawing, being especially good for long lines of uniform width, and they are excellent for stippling. The crow-quill pen, however, produces a lively, varied line better suited for indicating curves, shadow, and depth.

The ink drawing surface should be protected from finger smudges. Any oil, even a light touch, may cause ink lines or dots to spread or "fuzz." Use a slip of paper between your hand and the drawing, or make a mask to protect it (see p. 26).

Different artists develop their own particular ways of drawing; there are no fast rules to follow. Some methods of using the pen, however, give better results than others. If the pen is held too loosely or too tensely, if the pressure is greater on one nib than on the other, or if the point is at too acute an angle to the paper, the line will be ragged and blotchy. Firm but not excessive pressure and a wide angle between the pen and the paper (fig. 20) will help produce smooth, clean lines. A new pen point usually has a thin coat of oil that must be removed before ink will flow smoothly. Clean the new point with soap and water or rubber-cement thinner before using. The pen should be wiped on a chamois at frequent intervals while you are working, since the point tends to become gummy with drying ink. Dried ink can be removed by rubbing with an emery cloth or by cleaning the point with any good liquid household cleanser—wipe it dry or it may rust.

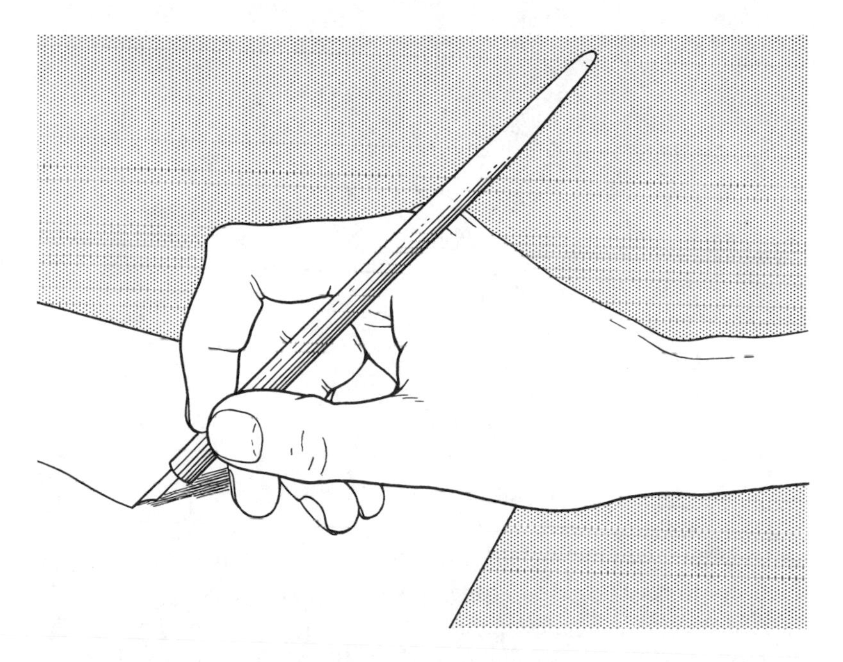

FIG. 20 Correct position of the hand in holding the drawing pen.

The pencil outline on the inking (final) ground is the basis for the ink rendering. In biological illustrations an outline generally separates the subject sharply from the background. The line should be varied in width, to add interest and give emphasis. This can best be accomplished by using a flexible point and varying the pressure. If no outline is to appear in the ink rendering, the shading lines or dots are drawn to the penciled outline (see fig. 25).

A line should begin at a natural starting place, such as a corner or the junction of two lines. If possible, do not lift the pen from the paper until the next junction or angle in the line. If you must break the drawn line at some inconvenient point, you should continue by leading into the inked line in the same direction, touching the pen to the line very lightly and carefully just before the break in the line (fig. 21). This must be done with a swinging motion or the point of junction will be obvious. A second method of continuing a line that has been broken is to

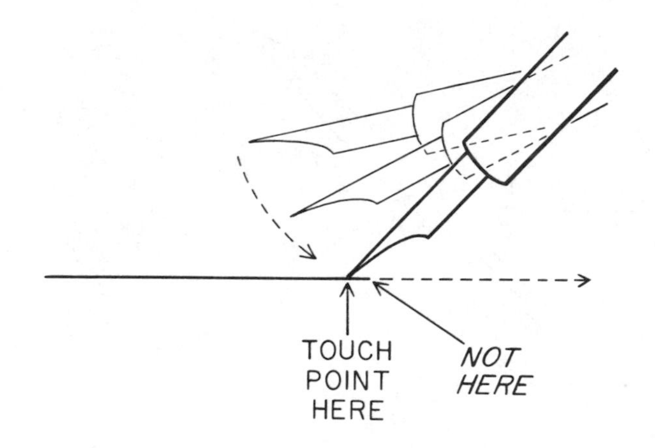

FIG. 21 Continuing a broken inked line.

begin the new part without touching the old line and then fill in the gap with a finer pen (fig. 22) after the previously inked lines dry.

The work should proceed gradually over the whole drawing, not in strips or areas; that is, detail and shading should be done layer by layer rather than piece by piece like a jigsaw puzzle. Attempting to finish one area at a time is likely to introduce errors in shading and emphasis. The completed area may be quite dark, and in trying to adjust the rest of the work to this area, you may make the whole picture too dark. If you squint at the drawing from time to time while working, you can more easily perceive general areas of value and contrast, and you will be able to correct and develop the shading before things get out of hand.

Another reason to proceed gradually over the entire picture is that an

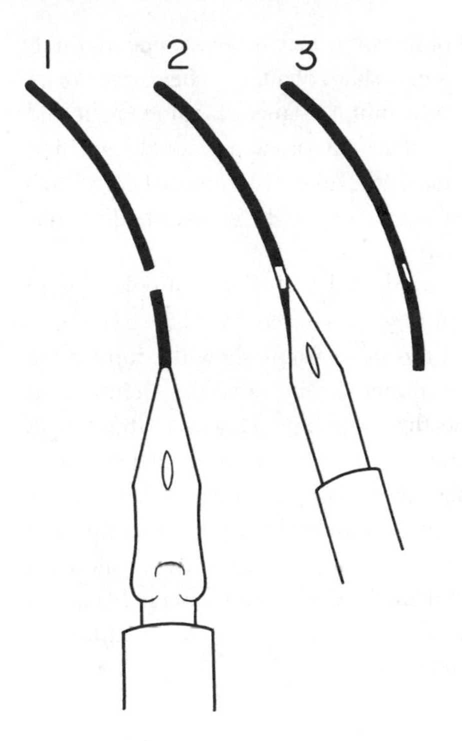

FIG. 22 Joining inked lines using a fine pen.

artist tends to draw differently at different times, which could cause the drawing to look disjointed.

It is wise to keep the drawing lighter in tone and to space the shading lines or dots a little wider than you really desire, because reduction and printing tend to thicken the lines slightly and "close up" the drawing, thus darkening the picture in general. By looking at the work through a reducing lens from time to time, you can better judge how much shading and detail to use.

Every drawing, even a simple outline, should show the direction of the light source. In an outline drawing you can make the lines heavier on the shadowed side of the figure than on the lighted side. More elaborate drawings are shaded more completely, as is discussed under the different techniques. The amount of light a surface receives depends on its angle in relation to the light source. A sharp change in the surface of an object will show a correspondingly abrupt change in value (light and dark), whereas a gently curved surface will show a gradual transition from light to dark. Shading is used to explain the object being illustrated; the artist must decide how much is enough, for overshading may obscure details.

A pen-and-ink drawing may be shaded with dots ("stipple"), with short lines ("hatching" or "hachure"), or with contour lines. Hachure lines that follow the contours of the subject may show the form more effectively than stipple in many instances, and once the technique is learned hatching takes less time than stippling. However, stippling is easier for the beginner to do, since it requires less care to achieve satisfactory results. There is no danger of hooking the short strokes (see below), and an imperfect dot is much less obvious than an imperfect hachure line. Contour lines require careful planning and execution but are the most effective and elegant method of shading. Scratchboard is especially suited to contour-line shading (see fig. 28); the technique is discussed in the section on scratchboard (see p. 52).

STIPPLING

Stippling requires a rigid pen point. If you use a crow-quill pen, choose a medium to large point and hold the pen vertically, not lifting it more than 2 or 3 millimeters from the paper. The size of the dots is determined by the size of the point, not by the amount of pressure used. Pressing the pen hard will make triangular stipple dots and spatter ink on the paper. When shading from light to dark, simply space the dots more closely; do not use bigger dots. A rough texture may be suggested by stippling that is uneven in both spacing and size of dots. Figure 23 is an example of a drawing shaded with stipple.

A technical pen (fig. 7) or the Leroy socket holder with the appropriate point (fig. 24) will make stippling not only easier but more even. These pens let you avoid making triangular stipples, and you can choose the size of the dots with ease.

In general, stipple dots should not touch one another, except where this is unavoidable in very darkly shaded areas. Stippling is done at random unless a pattern is desired for a special effect. Each stipple dot should be placed with care for its size, symmetry, and spatial relationship to other dots. A careless, "rapid-pecking" approach produces poor-quality stippling.

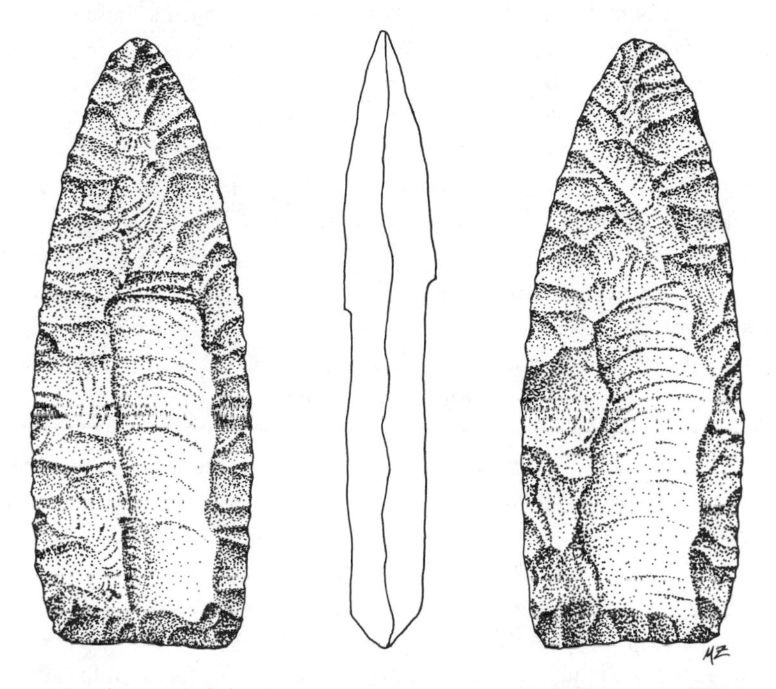

FIG. 23 A drawing shaded with stipple. The subject, a Clovis point replica, was drawn $1\frac{1}{2}$ times natural size so that a one-third reduction, as reproduced here, would depict the point at its actual size. (Courtesy Matthew K. Zweifel)

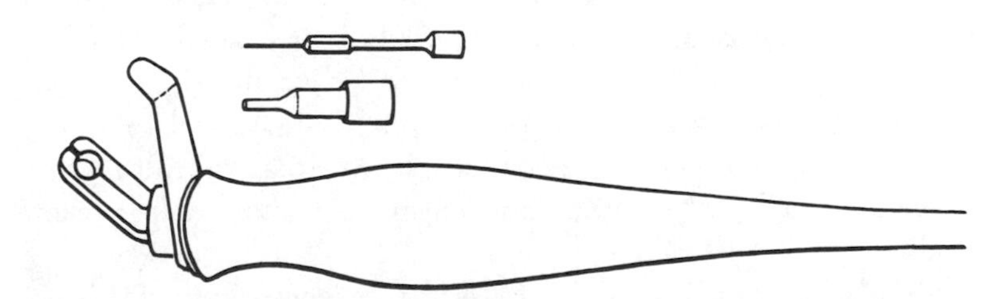

FIG. 24 Leroy socket holder and pen point.

Common faults in stippled drawings are dots so fine that they are lost in reproduction or so close that shadowed areas show up as solid black or as splotches of fused dots. To shade more deeply than close stippling allows, stipple closely and then join individual dots with a fine pen. The result should look like white stipples on a black background and can be graded into conventional stippling where a gradual change of tone is needed.

HATCHING

Stippling by hand, if carefully done, is extremely time consuming. It is worth the trouble to learn to use hachures for greater speed and convenience. In many instances a few well-placed hachure lines will show the contours of the subject satisfactorily where you would need several thousand stipple dots to achieve the same effect. In another instance it may be desirable to use both hachure and stipple in the same drawing to distinguish structures—skin from fur, for example, or cartilage from bone.

Hatching lines should run with the contours whenever possible. Use a flexible point so you can change the pressure while drawing and thus vary the line from fine to broad. The strokes should not end in "hooks"; pick up the pen at the end of the stroke and then move into the next stroke. Occasionally, however, hachures are slightly curved at the ends to suggest creases or indentations (see fig. 25). It may be advisable to plan each line lightly in pencil on the final drawing paper and to follow these guidelines exactly with the pen.

In some instances you may not want to vary the width of the hatched lines; instead, increase the number of hachures to give the impression of shadow, or use cross-hatching. When well done, cross-hatching can be very effective and labor saving; done carelessly, it may ruin the drawing. Cross-hatching should never be done at a 90° angle; this gives the mechanical look of graph paper. Cross-hatch at an acute angle, drawing tiny diamonds instead of squares. A difficulty with cross-hatching is that it is more likely to "close up" and blot in reproduction than plain hatching.

Because I assume that the work of the biological illustrator is intended for publication, the directions and suggestions I give here are directed toward that end. Often the printer hears an illustrator say it is all right if particular fine lines or fine stipple dots "fall out" during reproduction, because "they are really not essential anyway." The artist supposedly has drawn each line and dot to give the viewer a certain impression. If the absence of such lines or dots would not alter that impression, they should not be drawn in the first place. The camera cannot choose what to keep and what to drop out; it tends to retain only parts of too-fine areas, perhaps creating an impression somewhat different than

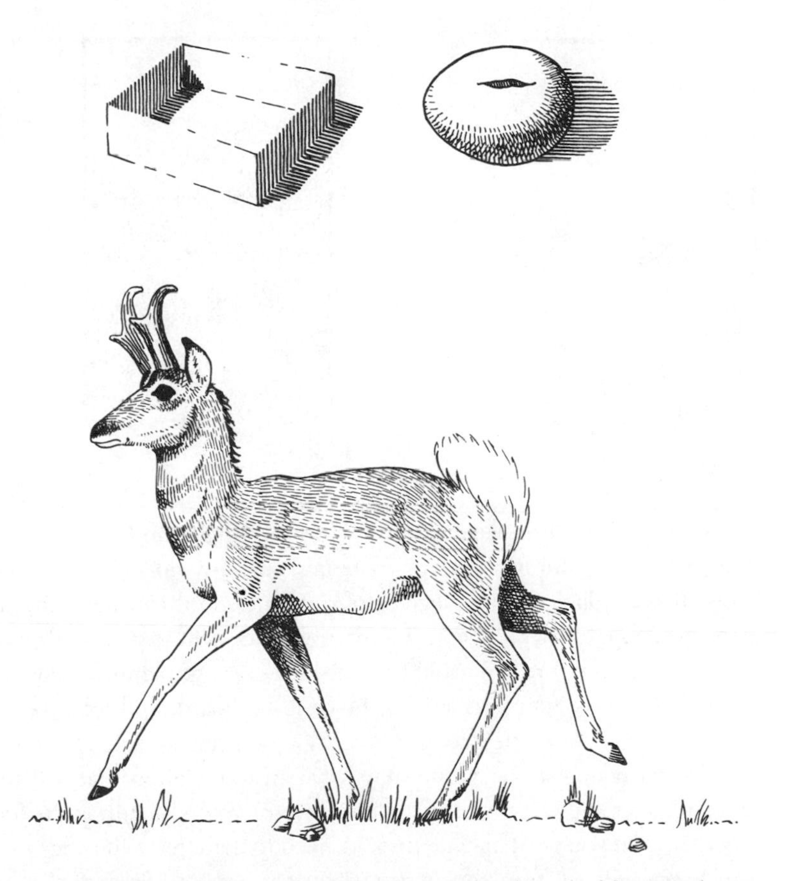

FIG. 25 Examples of hatching and cross-hatching. Reduced one-third from pen drawing of the pronghorn, *Antilocapra americana*.

intended. If all lines and dots are to be reproduced, you must draw them firmly, which does not mean coarsely. If in doubt about a line, strengthen it or eliminate it.

SPECIAL TEXTURED GROUNDS

By using a textured ground, an illustrator can produce a satisfactory picture in a fraction of the time it would require to stipple the same area by hand (see fig. 3).

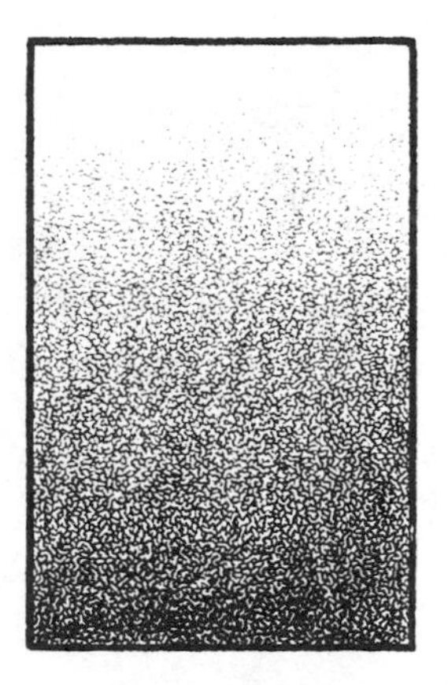

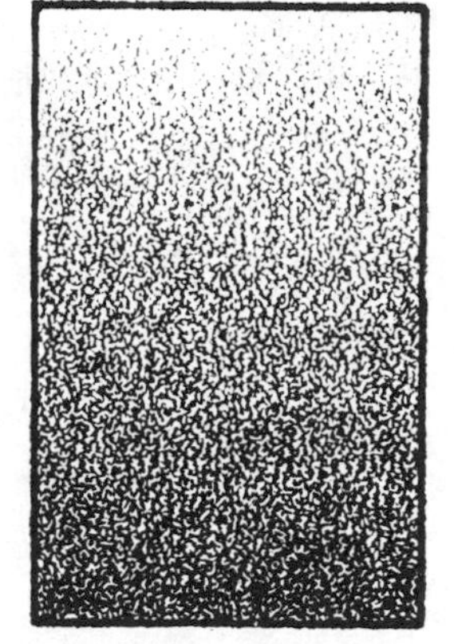

FIG. 26 Examples of coquille board.

Coquille board is a paper ground whose textured surface resembles finely wrinkled paint or silk crepe material, with tiny ridges and depressions. It is available in two degrees of roughness and can be found at large art-supply stores. The textured surface is delicate and will not stand vigorous erasure. To avoid the need for erasing, plan the drawing completely before transferring it to the coquille board, and refer to the sketch while shading the drawing. Use as little pressure as possible on the transfer pencil; a heavy line may leave an indentation that will refuse pigment and show up as a white line in reproduction. (On the other hand, if you want a white line, or need to demarcate the edge of a toned area without using an outline, you can emboss, or indent, the line or edge with a stylus before you apply a shading medium.)

For black-and-white reproduction, the drawing is done with medium and soft lithographic (black wax) crayons or black "all surface" pencils. (Graphite pencils or charcoal pencils may also be used, but such renderings may have to be reproduced as continuous-tone illustrations; see pencil drawings, p. 59.) You can outline in ink and shade up to the outline with the crayon, draw the outline lightly in crayon, or use no outline at all (you can emboss the edge of the shaded area, as suggested above). It is possible to stipple with ink in the darkest areas, blending the

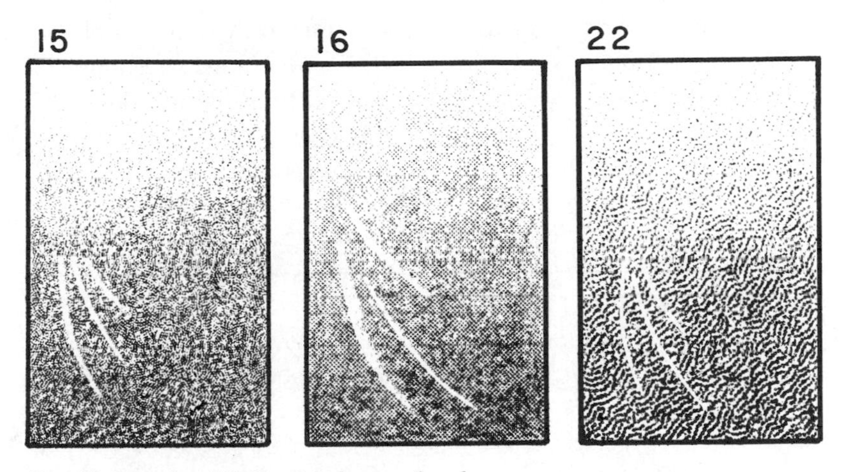

FIG. 27 Examples of EssDee British scraperboard.

stipple into the crayon shading. Apply the pencil or crayon carefully to avoid definite lines that cannot be altered on the drawing. Use a circular motion, gradually building up the tone. Rubbing the surface of coquille board lightly with the crayon produces a fine random stipple effect, and harder rubbing produces a darker, more closed-up effect (fig. 26). Fuzzy or ragged edges are smoothed after the drawing is completed by painting with white gouache such as Pelikan or Pro White. It is possible to lighten an area that is too heavily shaded by stippling with white gouache paint.

Another special ground is a scratchboard with a textured surface, called EssDee suede-finish scraperboard. It is not readily available but can be ordered (see Selected References). It is used in the manner of coquille board but has the advantage that it can be cut, or scratched, with a knife or scratchboard tool, greatly increasing the illustrative possibilities (fig. 27).

These special grounds indeed save time and labor, and illustrations done on them can be superb. But they are considerably more expensive than materials for simple pen-and-ink drawings and also are harder to obtain. In addition they are fragile, so they must be handled with care and mounted on a firm backing.

SCRATCHBOARD

Scratchboard is a thin cardboard coated on one side with a layer of fine chalk. Ink is painted or drawn on the chalk surface, and scratching the ink with a sharp tool exposes the white ground. Scratchboard also comes already blacked, and the entire drawing is scratched out.

There are several brands of scratchboard. In general, the chalk layer of the cheaper brands is too soft and thin. The coating on more expensive brands may be thicker but gritty. Before buying, carefully inspect each sheet for imperfections such as tiny cracks that will hold ink too deeply to be removed.

Use a razor to trim scratchboard to a convenient size; scissors may cause cracks along the cut edges. Allow a good margin, at least 1½ inches, all around the drawing, and securely mount the scratchboard on a slightly larger piece of posterboard before beginning to draw. A scratchboard tool that fits into a penholder is available from art-supply stores, but an X-Acto knife with a number 11 or 16 blade is excellent.

It is important to protect scratchboard from being smudged with oil from your hands. The slightest trace of oil on the ground will cause subsequently drawn lines to "fuzz." Keep a clean slip of paper between your hand and the drawing surface or protect the drawing area with a mask (see p. 26).

Before beginning to draw on scratchboard, be sure it is thoroughly dry. Placing it on a warm radiator or under a light bulb for a few minutes may be sufficient; some illustrators find a hand-held hair dryer invaluable in damp weather for drying other water-based media as well as scratchboard.

The scratchboard surface should be prepared before you begin any drawing. Erase carefully and thoroughly with an Artgum ("soap") or a

Pink Pearl eraser, and wipe the surface clean with a soft cloth. Burnish the board with pounce, or draftsman's pumice, and wipe this off smoothly.

There are two usual methods of working on white scratchboard. In the first method, a silhouette of the figure is painted black. (Ink straight from the bottle may tend to clot; dilute it slightly and apply several washes of this solution, allowing the surface to dry between applications.) After the surface is thoroughly dry, the artist scratches (or scrapes) white lines and areas into the black silhouette.

In the second method, the figure is drawn lightly on the scratchboard with a medium-lead pencil. Contour lines and deeply shadowed areas are carefully planned in pencil and then are drawn in ink. These inked lines then are scraped, to shape and thin them as desired (see fig. 28A,B,C). Either a fine brush or a pen may be used. The pen will produce a dense black line but may become gummy or gouge the chalk surface. The brush, then, is preferred for drawing on scratchboard, but the lines must be drawn densely black, not allowed to thin to gray. Ink must be applied smoothly, not in gobs or puddles, since these will affect the drawing surface.

Allow all ink to dry thoroughly—if you attempt to scratch while the ink is still damp, the chalk layer will crumble away like cheese. Blowing on the surface to remove dust tends to dampen the chalk layer. In humid weather the scratchboard may have to be dried every few minutes.

You can buy scratchboard already blacked. To transfer a sketch to black scratchboard, make a "carbon" on the back of the sketch with soft-lead white pencil and draw over the lines of the sketch with a 4H pencil, as directed on page 40.

Carefully plan the lines on the silhouette with the point of the scratch knife. Then very gently scrape them to the width and length you desire. Figure 28 demonstrates successive steps in scratchboard technique, from placing the lines to completing the shading. Usually finer lines are used to suggest distance, whereas wider white and black lines are used for the nearer parts of the subject. In a given area the black lines are thin and widely spaced in the light region and become closer and thicker as the shadow increases.

Rather than actually scratching the board, use a light scraping mo-

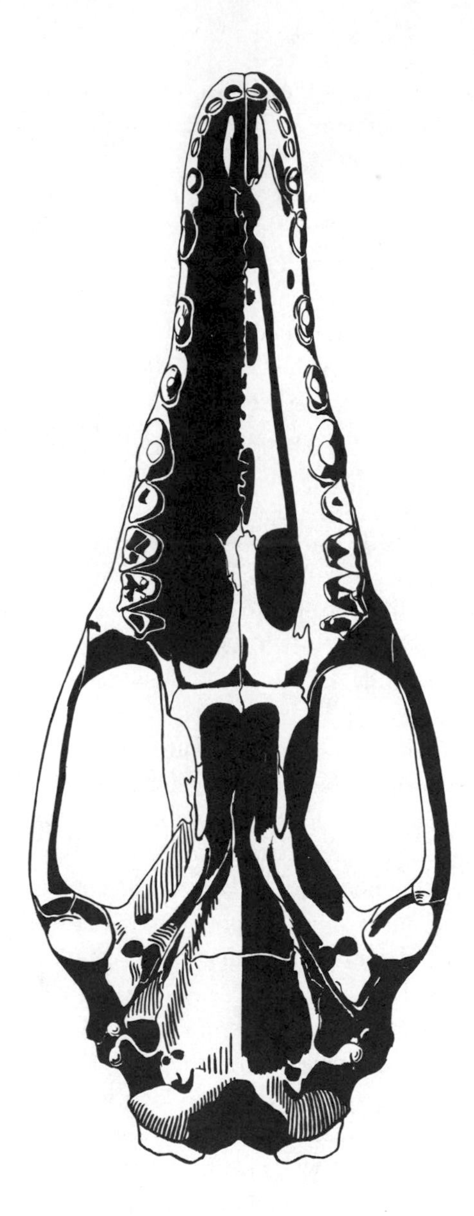

FIG. 28 Steps in making a contour-line drawing on scratchboard. Reduced slightly from a drawing of the ventral view of the skull of the bandicoot, *Peroryctes broadbenti*. A, The outline is drawn, and all shadowed areas are painted black. (Some hatching is drawn with a pen, to be scratched later.)

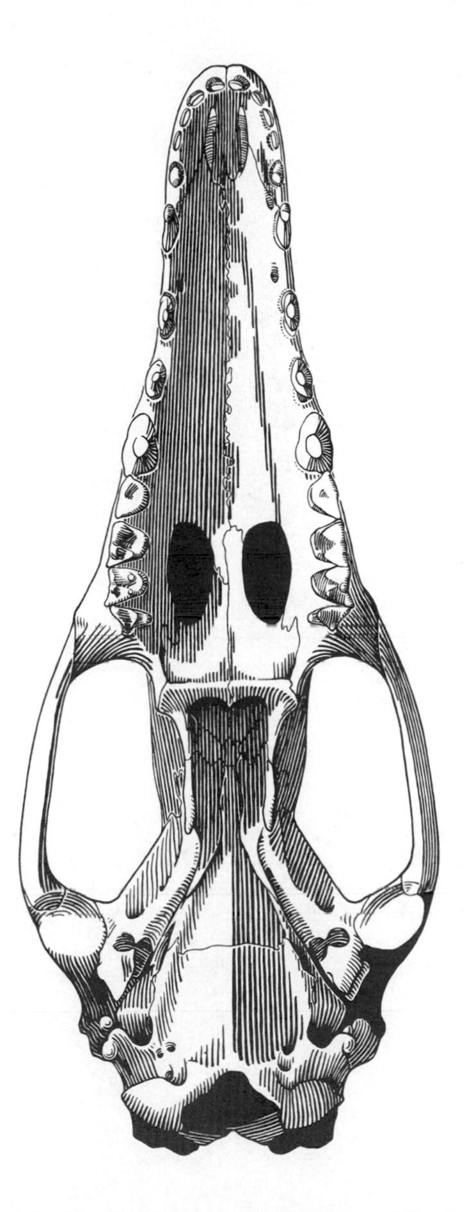

FIG. 28 *continued* B, The hatching (contour lines) is carefully planned with a fine X-Acto knife. Distant parts are separated from foreground parts by breaking lines at points of junction. Sutures are drawn through hachure lines. Modeling of contour lines proceeds gradually.

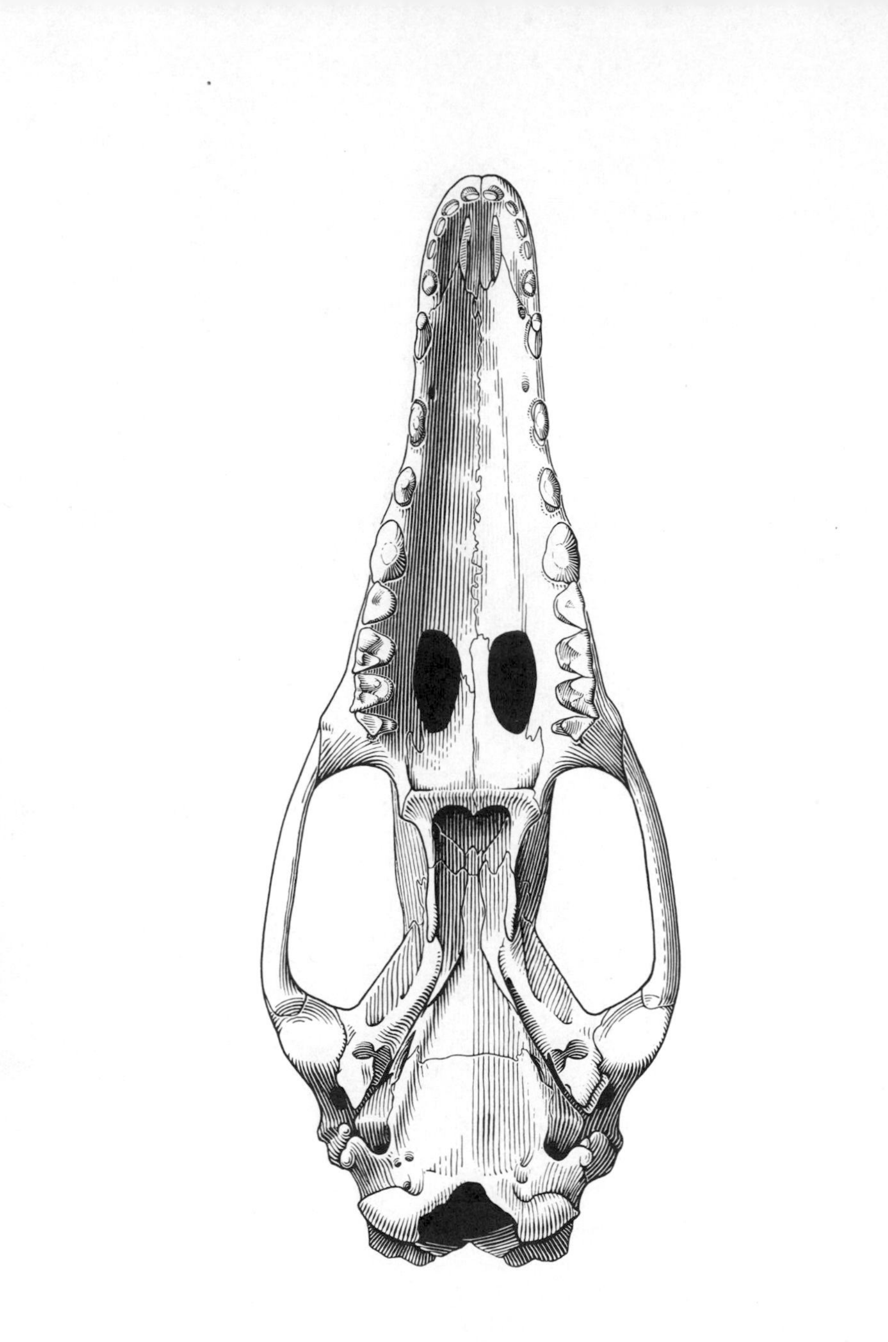

FIG. 28 *continued* C, Suture lines are brought into prominence by breaking the hachures in contact. The shadows are adjusted by carving each contour line.

tion that will remove ink with a minimum of gouging and routing of the chalk surface. If you scrape gently, the area can be reinked and re-scratched. Too much reworking will eventually remove the chalk down to the cardboard backing.

A common mistake when using scratchboard is to scrape the lines too fine. Extremely fine lines may not reproduce at all, and dark areas with very fine white lines may close up in reduction. In general, fine broken lines are unsafe, since they are frequently lost in reproduction.

The advantages of scratchboard over pen-and-ink drawings on ordinary drawing board are several: long, thin lines may be done with ease; an unsatisfactory line may be redrawn; the shading can be adjusted as the drawing proceeds and may even be changed if desired; mistakes and ink blots are simply scraped away. These qualities make scratchboard an excellent ground for simple outline drawings (figs. 13, 29) as well as for more complex renderings. Subtle shading and use of contour lines are possible with scratchboard, and it is incomparable for drawing fur.

Scratchboard is far more expensive than ordinary drawing board. Fortunately there is a satisfactory substitute, at lease for simpler drawings—frosted acetate film. This film is sold under several names, in various thicknesses, in flat tablets and also in rolls (less desirable because they tend to curl up again). This ground is flexible and is not delicate like scratchboard, but it must be mounted on a white background for use. Because the film is transparent, the preliminary sketch can be traced directly, eliminating the transfer procedure.

There are special inks for drawing on film, but most waterproof drawing inks work well. The Koh-I-Noor Rapidograph "imbibed" eraser, made especially for these grounds, will erase ink without damaging the surface of the film, and the area may then be reinked.

The illustrator draws with pen or brush on the frosted side of the film. Lines may be scraped, as on scratchboard, but the thin film cannot be reinked and scratched over again. Film should not be placed near heat to dry the ink. As with other drawing surfaces, the ground should be protected from contact with skin oil while working.

The view from the nonfrosted side of the film will appear sharper (and of course in mirror image, which can be corrected in printing by "flopping" the negative), but either side will reproduce well. If necessary, a drawing may be washed off with soap and water and the film reinked as long as the surface is unscratched. You can even wash off small areas with a moistened cotton swab, using a shield to protect the rest of the work and wiping away from the shield.

In time acetate films may shrink slightly; in humid conditions they may swell. For most purposes this will not be important. If it is necessary to retain the finished measurements absolutely, another flexible ground is available—polyester film (often called Mylar). Polyester film

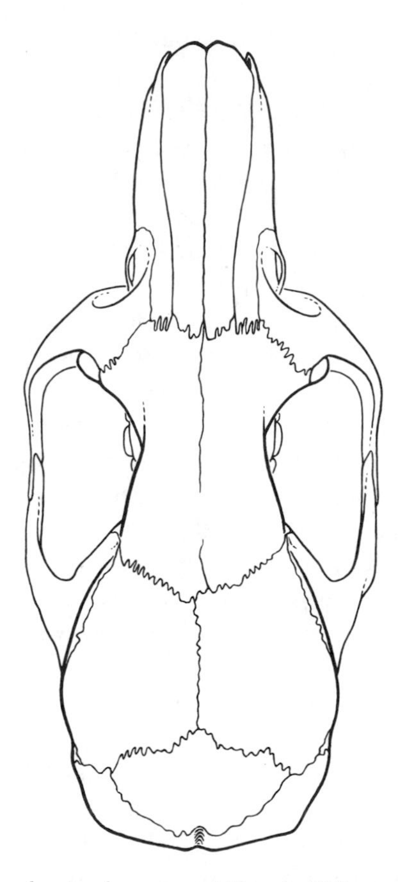

FIG. 29 A simple line drawing done on scratchboard. All lines are drawn with a pen, then scraped to the desired thinness. Reduced one-fourth from original drawing of the skull of the giant rat, *Xenuromys barbatus*.

takes ink well, but scratching may leave an unattractive gray patch. Also the surface may be harder and less easy to scrape smoothly, so that the lines are not as clean as those scratched on acetate film. Test the film before you buy, or buy only a small quantity at first.

Continuous-Tone Illustrations

Artwork that contains shading in tones of gray, such as pencil drawings or wash drawings, is called *continuous-tone* copy (sometimes, erroneously, halftone). Like photographs, continuous-tone artwork must be reproduced by the halftone process, using a screen that divides the tones into tiny dots of varying sizes. This technique is more critical than reproducing line copy, so continuous-tone illustrations should undergo less reduction in reproduction. In general, plan for no more than one-third off or a 33⅓% reduction. Figures 1, 11, 30, and 59, among others, are continuous-tone illustrations.

PENCIL DRAWINGS

You will need pencils ranging in hardness from 4H down to 6B (for the darkest shadows). The ground may be any smooth-surface paper that has enough tooth to hold the graphite without showing an obvious grain. Kid-finish bristol board, plate-finish drawing paper, and frosted films are all possibilities. A sandpaper block is useful for keeping the pencils sharp. A kneaded eraser, a vinyl eraser sharpened to a point, and several rolled-paper smudgers, or tortillons, are also needed. These last can be bought at art-supply stores or made by rolling a strip of paper tightly with one end protruding in a point (fig. 8). Some artists use a finger to smudge or blend the shadows, but this mixes oil with the graphite and makes erasure difficult. Use a slip of paper under your drawing hand to keep the ground clean or, better yet, mask the paper except for the work area (see p. 26). Tape the mask in place so it will not slip and smudge the drawing.

Plate-finish paper is used when very precise, tiny detail is required. Since this slick surface does not hold graphite as well as a rougher surface, use harder pencils. Keep in mind that a softer lead may be difficult to apply over a hard lead.

The outline and preliminary lines are drawn lightly with a sharp medium-to-hard pencil. Use softer leads to draw in the shadows, the softest leads in the darkest areas. The difficult part is putting down the strokes without showing obvious lines. By holding the pencil almost horizontal and stroking it back and forth, you will get an even tone. Softer leads and more stroking will produce darker shadow. Shadows may be hatched and the hatching smudged, or blended, for a softer, more natural effect. Use the smudger lightly; do not grind in the graphite, just stroke it. Hard leads do not smudge easily and so are likely to remain as obvious lines.

The kneaded eraser can be used to clean around the drawing or pinched to a point to blend small areas of graphite. It can also lighten a wider area by "picking up" excess graphite: flatten the eraser and press it

FIG. 30 Example of pencil technique. Reduced about one-third from original drawing, on vellum-finish bristol board, of seedpod of a desert plant, *Proboscidea* sp.

gently against the area of the drawing to be lightened. The sharpened vinyl eraser is used to "lift" highlights in points or thin lines.

Bright highlights should be planned for and left untouched in the working drawing. If, after you have completed the drawing, you decide you need a highlight, there are a few emergency techniques. You can scrape, with care and a light touch, a bit of the ground, or apply a touch of white gouache paint. (After you use these techniques, no more penciling should be attempted.) It is easy for scraping to get out of hand; use extreme caution. Likewise, use white gouache as little as possible, since it attracts pencil dust and can flake off.

Pencil drawings are delicate and should be protected against accidental smudging. There are spray fixatives, but these must be applied carefully, lightly, and evenly lest they cause the graphite to puddle or run. Always apply the fixative to a test sketch first, even if you have used that particular can safely: different grounds react differently, age and temperature may affect the spray, or the valve may have been damaged. Also, protect yourself from toxic fumes; use all sprays in a well-ventilated area.

WASH DRAWINGS

In wash drawings, as in pencil drawings, the white of the illustration is the white ground (or white background as seen through a transparent ground). The dark is the shading material that is applied to this white ground (fig. 31).

The most convenient ground for wash drawings is illustration board, available in kid and smooth finishes. The board can be handled easily, moved and tilted as desired. The paint should be a good-quality, permanent black watercolor in a tube, such as Winsor & Newton. Beware of poor-quality or old paints; they may not dissolve completely, making it impossible to apply an even wash. You will need several watercolor brushes, flat and round (fig. 32). A wide, flat brush or a large round brush is used to apply the preliminary clear water wash and the broad areas of pigment wash. Smaller round brushes are used for details. You will also need a white china saucer or a purchased palette and two one-pint containers of clear water.

Be sure the drawing surface is absolutely clean; every dust speck or eraser crumb or tiny hair must be removed. You may need to erase the whole area with a kneaded or gum eraser before beginning to work and thoroughly brush away all eraser crumbs.

If you are unfamiliar with this technique, you will need to practice making a smooth wash before you attempt an illustration for publication.

A flat wash is applied in the following manner. Squeeze a dab of black paint into the saucer. After drawing several squares about 2 inches wide on a practice board, wet one of the squares with clear water and let

FIG. 31 Example of wash technique. Reduced about one-third from original drawing on kid-finish illustration board, of the frog *Phrynomantis slateri*. (From R. Zweifel, *Bulletin of the American Museum of Natural History* 148, art. 3 [1972]: 411–546)

it dry just until the water loses its shine. Tilt the board slightly toward you. Now mix a touch of paint with water, enough to cover the wet square. Load the brush and, starting at a top corner, paint a strip from one side to the other; without lifting the brush, move down and paint back across the square, slightly overlapping the previous line. Repeat this until the square is covered. You may have to reload the brush on the way down. Always move from top to bottom, back and forth, lightly

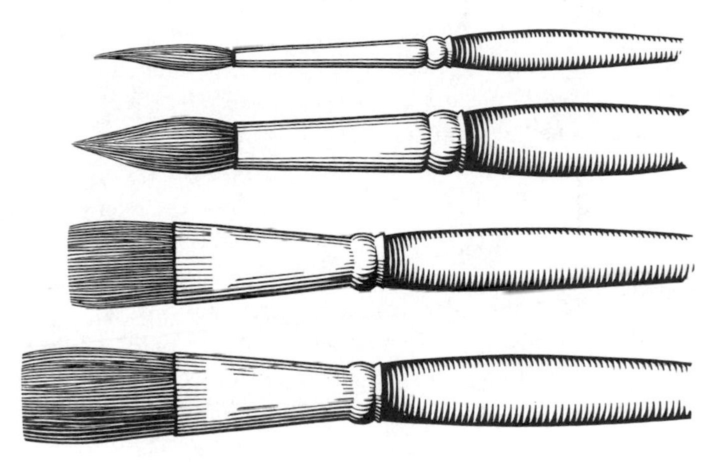

FIG. 32 Flat and round brushes.

without scrubbing. The last stroke may be applied with a less wet brush, and the tiny puddle of surplus pigment at the bottom can be absorbed with the squeezed-out brush. The board can be tilted as desired to aid or retard the flow. If done properly, the whole area should be covered with an even layer of pigment. Try pale, medium, and dark washes. (The wash will be lighter when dry.) The paper should be less damp to receive a darker wash and almost dry for small dark areas.

After an area is completely dry, you can wet it again with clear water and apply another wash over the first. Because the paint is permanent, you can build up layers of shading in this manner, but it is not possible to remove pigment once it is down. It is best, therefore, to start the illustration with the lightest wash you will need.

Once the flat wash is mastered, try washes darker at top or bottom, grading them smoothly. Then apply graded washes to outline sketches of cubes, cylinders, circles, and cones—the basic shapes in nature. Keep in mind the shading convention, with light coming from the upper left of the picture.

To do a wash illustration, transfer the sketch lightly to the cleaned illustration board. A mask may be cut to shield the drawing area from oil and smudging (see p. 26). Wet the figure area with clear water and apply a light wash to the entire figure. Then proceed to build up areas of shadow. In general a lighter-toned picture is more pleasing than a darker one, but the range of values should be sufficient to represent the subject well and to make the drawing interesting.

Areas to be left white may be painted with a purchased liquid mask before you apply the first light wash ("paint" the wet brush across a bar of soap before dipping it into the liquid mask for easier cleaning); after the wash drawing has dried thoroughly, rub off the mask with a bit of kneaded eraser. Or you can add highlights with white gouache paint after the painting is dry. (Remember that pure white areas must be handled specially by the printer, adding to the cost of publication.)

Stippling and line work done with pen or brush must be added after the paper has dried. Solid lines will appear broken in halftone reproduction and should be avoided unless a fine screen is to be used.

Wash drawing is an excellent technique for rendering qualities such as softness and furriness and certain light-and-shadow effects. Once the wash technique is learned, a drawing can be done quickly. Mastering this medium, however, takes time and practice; it is no easy matter to achieve a finely graded wash or to apply tones without blurring the edges. The excellent results are well worth the effort.

CARBON-DUST DRAWINGS

The carbon-dust technique, which has come into frequent use over the past several years, combines pencil drawing and painting. As with wash drawings, it takes practice to master the technique, but work of great delicacy, detail, and softness of shading can result. See figure 1 for an example of a carbon-dust drawing.

The ground for a carbon-dust drawing must have enough "tooth" to hold the powdered carbon but should be smooth enough so the powder can be wiped and blended. Almost any fine-quality surface can be used, such as vellum bristol, the best-quality scratchboard, and frosted acetate or polyester film.

The carbon dust itself is ground from charcoal pencils of varying hardness—H, HB, and B. You will need a flat dish to hold the powdered carbon. To "paint" the dust on the ground, you will need several brushes: a round or flat watercolor brush of the size used for painting wide washes, one or two smaller soft-bristle flat brushes, and some fine round and flat brushes for putting in details. (Use good-quality brushes, and save them for this technique only.)

Additional materials required are a sandpaper block, small pieces of clean chamois, and some sharp-pointed rolled-paper smudgers, as well as a vinyl eraser that can be sharpened to a wedge or a point and a kneaded eraser. An X-Acto knife may also be useful.

One problem with carbon dust is keeping the work area clean. The dust is so light that it tends to drift about, smudging every surface it touches. A prudent move would be to cover the work area with a large sheet of plain paper, securely taped down, that can be rolled up and discarded when the drawing is completed. Another precaution: don't sneeze.

The drawing surface must be absolutely clean, free from all specks of dust, eraser crumbs, hairs, and skin oil; any irregularities, such as scratches and fine indentations, will show to disadvantage. Erase the ground with the kneaded eraser, wipe it with a soft cloth, and handle it only by the edges. If you are using a flexible film, tape it all around its edges to a firm board, for ease of handling and to prevent carbon dust from working its way underneath. A mask may be used to protect the drawing area while you work.

The preliminary sketch should indicate all the necessary shading. If a transparent film is to be used for the finished carbon-dust drawing, spray the preliminary sketch with fixative before tracing so it will not transfer to the back of the film (or make a photocopy of the sketch and trace that). If the sketch must be transferred to an opaque ground, use a very light touch so as not to indent the drawing surface.

Make a supply of carbon dust by rubbing one of the pencils on the sandpaper and catching the dust in the little dish. Make a supply of each hardness; about a quarter of a teaspoon is enough to start with. The dish should have a cover so the dust will scatter as little as possible.

Have ready a scrap of the chosen ground on which you can try out strokes and densities of carbon dust. This will also be useful later for experimenting with fixatives.

Once the sketch is transferred or is in place to be traced, outline the figure by drawing as lightly as possible with the HB carbon pencil. (An outline may not be desirable; if you find it necessary, you can add one later.)

Begin to apply the carbon dust. With the largest brush, pick up a dip of dust, tap the brush to remove any excess, and "paint" the carbon on the drawing with soft, sweeping strokes. Use as light a touch as possible; do not grind the dust into the drawing surface. Apply dust to the entire drawing, gradually layering to obtain darker areas. Establish the large areas of shadows and lights first, as described above; details will be worked later with the smaller brushes.

A tiny roll of chamois can be used to blend the dust smoothly. Use a fold of the chamois to blend larger areas. The rolled-paper smudger also will blend, or it can be used to paint in darker tones—always tap to remove excess dust before touching brush or smudger to ground. Darkest areas may be penciled and then blended with a smudger, and this action can be repeated, working up layers of tone.

Pick up lights with a bit of rolled chamois and the kneaded eraser. By pinching and shaping the eraser, you will be able to lighten very small areas. Sharpen the vinyl eraser and pick out highlights; cut it into a wedge to erase a fine line. The edge of a piece of blotting paper also will remove a thin line or will draw one in if used to apply the dust.

After the carbon-dust drawing has been completed as far as possible, you may put in the finishing touches. Draw darkest darks in india ink or soft graphite pencil. Use the X-Acto knife to scrape fine highlights or lines, or apply a touch of white pencil or gouache paint. (Once the surface has been altered by one of these techniques, no more carbon dust can be applied.)

Clean the drawing with the kneaded eraser. Rubber-cement solvent may be useful to remove fingerprints, but test it first on the scrap of ground you used to try out strokes.

Carbon-dust drawings are delicate and must be protected. Before spraying fixative, test it on the scrap—some fixatives melt some grounds, and some spot carbon dust. Mount flexible grounds firmly and mat the drawing with a thick mat. Cover the mounted work with a cover paper taped to the back and folded over the front. Tape this at the bottom to keep it from touching the drawing surface.

Carbon-dust drawings must be reproduced by the halftone process, like wash drawings and photographs, and the same cautions apply.

Methods of Correcting Mistakes

There are several methods of erasing or eradicating errors in ink. Scraping with a sharp knife or razor may remove ink, but unless done with extreme care it will remove the paper as well. A fiberglass eraser will do the same. Rubbing a piece of soapstone over the roughened surface may smooth the damaged area enough so you can redraw the erased line. (Soapstone can be purchased at art- and sculpture-supply stores and in hardware stores.) The Koh-I-Noor Rapidograph "imbibed" eraser will erase most drawing inks from acetate and polyester films, and even some tracing vellum, without damaging the drawing surface. An electric eraser is excellent for use on inked drawings and is a worthwhile purchase if you intend to do much illustrating.

Erasure is not always advisable or necessary in line drawings, as in the event of large blots of ink or accidental lines that do not mar the lines of the drawing itself or that appear in areas with no drawing. This type of error can be painted over with superwhite gouache or covered with a strip of white paper or white artist's tape. These cover-up methods may also be used to thin lines in the drawing and to straighten blotted or blurred edges of inked lines. Among the best white-out products are Pelikan Graphic White, Dr. Martin's Bleed Proof White, and Pro White. Pencil lines should be removed from finished ink drawings in a way that will not lighten or smear the ink. A kneaded eraser is safest.

Errors and smudges on pencil drawings should be erased very carefully with a kneaded eraser. Errors in wash drawings cannot be satisfactorily erased, but smudges and fingerprints should be erased or cautiously painted out. Pasted strips of paper should not be used to cover errors on continuous-tone drawings, since the edges of the cover strip will be apparent in the reproduction unless removed by the printer, at extra cost.

6

Preparing Graphs and Maps

One of the most important tasks of the scientific illustrator is preparing graphs and maps. A properly prepared illustration of this type can convey a wealth of information quickly and accurately, often more effectively than is possible in the accompanying text.

Some cautions should be mentioned at the outset. Uneven freehand lettering detracts greatly from the professional look of the finished illustration; unless the artist is expert at freehand lettering, all lettering should be done with a lettering device, with commercial press-on lettering, or at least with the aid of a stencil. Phototype or computer-generated type may be useful, and you may be fortunate enough to find a typesetting device available at a large university or other institution (see p. 101).

Pay particular attention to reduction of the drawing for publication. Before planning any map or graph, you should know what publication the work is to appear in; ask for editorial guidelines well in advance. Many an otherwise good graph or map has been made virtually useless by reduction so great that important features were rendered illegible (see fig. 4).

Be certain that letters and symbols are large enough in the original copy. Do not assume that the editor will specify reduction appropriate to the size of letters and symbols. (If there is any question of this, mark the desired reduction on the copy in blue pencil.) Also, guard against trying to present too much information in a single figure, lest you lose the basic advantage of a graph or map—presentation of a clear, coherent picture of data.

Materials and Special Techniques

Here I consider materials and instruments especially suited to the construction of graphic illustrations.

GROUNDS

Graph paper is available ruled in either the metric or the English system. The lines may be printed in various colors: red, brown, black, green, blue, or violet. The choice of color is determined by whether it is desirable to have the complete grid system appear in the published figure. Pale blue usually does not reproduce photographically, so a figure may be submitted for publication drawn on pale-blue-lined graph paper without the grids showing in the reproduction. You can buy graph paper on which the printed grid is guaranteed to drop out.

A transparent ground is by far the most convenient for drawing maps and graphs. Tracing vellum is excellent for this purpose, as are acetate and polyester frosted films. The map or graph can be planned on any graph paper and simply traced without the problems of transferring. Acetate film may shrink or swell slightly, however; if the illustration must be kept for any length of time, it would be better to draw it on polyester film or tracing vellum. These flexible grounds should be mounted on a stiff white board for handling and mailing.

PENS

Technical pens (fig. 7) and the Leroy socket-holder pen (fig. 24) are convenient for lettering and for drawing lines. They are equally well adapted for use with a rule (curved or straight), a template, or freehand and will produce even-width lines, with width determined by the choice of point. A draftsman's ruling pen is more likely to cause blots at the beginning or end of a ruled line.

Several brands of technical pens are available, some fed by ink cartridges and others with built-in reservoirs.

If you want a line of varying width, use a conventional drawing—or crow-quill—pen. It may occasionally be necessary to darken a drawn line; this can be done by going over the dry line with a fine mapping point or fine-point technical pen, flooding the drawn line without widening it.

Technical pens invariably get clogged. They require cautious cleaning, since the points are fragile and costly. The first step in cleaning is to unscrew the point assembly from the rest of the pen and hold it under cool running water. If it is not too badly clogged, you may be able to flush out the point this way. Shake it gently up and down and listen to hear whether the inner parts are moving freely. Blot the point thoroughly to remove as much water as possible.

If your pen is more thoroughly clogged, flush out as much ink as you can with running water and then soak the point in commercial pen-cleaning fluid or in a solution of ammonia and water or ammonia-based detergent and water (soap or detergent is not adequate—ammonia is needed to dissolve these inks). Soaking the point for several hours or overnight may be necessary to dissolve the dried ink, but too long a soaking may damage the point parts. Rinse the point well and shake it gently to determine whether the inner parts are moving freely.

I strongly advise against taking apart the point itself, since the fine point wire is so easily bent; even a slight bend may mean buying a whole new point. If the point is hopelessly clogged, however, and you must either clean it or discard it, use extreme caution when taking it apart. Rinse all the parts free of flakes and chunks of ink, then soak them in cleaning solution. A soft toothbrush can be used to clean out grooves and threads. Rinse and dry the parts before reassembling the point. Do not try to wipe off the fine wire; replace it in the point assembly very carefully, using steady hands and a magnifying glass.

Art-supply stores sell "ultrasonic" pen cleaners that will quickly clean a point without your taking it apart if it is not too badly clogged.

INK

Use absolutely black waterproof drawing ink. There are special inks for technical pens, some specialized even to the size of point. There are also inks made specifically for different grounds such as acetate, paper, or cloth. One of the less troublesome brands of ink (in terms of clogging technical pens) seems to be Pelikan, but you should experiment to find what brand you prefer.

RULES

An 18- or 24-inch metal rule is essential. A metal rule generally is a truer straightedge than a wooden one, and the extra length helps in ruling long lines. Metal rules are available with "cork" backing; these are best because they slip out of position on the copy far less easily.

A wooden rule with a metal edge should be checked for straightness. Plastic rules may be straight enough but can warp with time. A clear plastic rule scored in a $\frac{1}{8}$-inch grid is valuable for making preliminary pencil drawings, especially right angles and parallel lines. Unless it has a beveled edge, any rule is likely to smear an inked line (see p. 76 for constructing a raised edge).

A T-square, used with a drawing board with two truly right-angle edges, will aid in ruling parallel and perpendicular lines. Clear-plastic drafting triangles, large and small, are useful for many purposes.

Curved lines can be ruled with a set of ship's curves or french curves (fig. 9). A flexible curved rule (fig. 10) may be helpful for ruling relatively long-radius curves, but is not useful in drawing tighter curves. You can construct your own short-radius curved rule by the following method: Draw the desired curve freehand on thin paper (tracing paper, bond, etc.). Paste this to a piece of mounting board or other thin cardboard and cut out the drawn curve with scissors or a razor blade. Smooth the cut edge with fine sandpaper or an emery board. Elevate this handmade curved rule from the drawing surface with a thin cardboard riser (p. 76, fig. 40).

RULING PARALLEL LINES

It is sometimes necessary to produce parallel lines as, for example, in placing grid coordinates or borders on a graph or map. Four methods described here will serve in most situations.

If a T-square and drawing board are available, simply mark off ticks along a line drawn parallel to one edge of the board. Slide the T-square from mark to mark, ruling as you go (work from top to bottom to avoid smearing the ink). A drafting triangle can be used in this same manner.

In the absence of a T-square, you can draw a line parallel to a given line by using a compass. Fix a compass point at any two points along the first line. Draw two arcs on a radius equal to the desired separation of the lines to be ruled. Draw the second line tangent to the two arcs (fig. 33).

At times you may want to rule evenly spaced parallel lines without the bother of measuring out odd fractions. Mark the desired rectangular space, as rectangle *ABCD* in figure 34. In the example, the rectangle is to be divided into eleven equal parts. Extend lines *AB* and *DC* as *EAB* and *DCF*. After finding any eleven convenient divisions, such as eleven

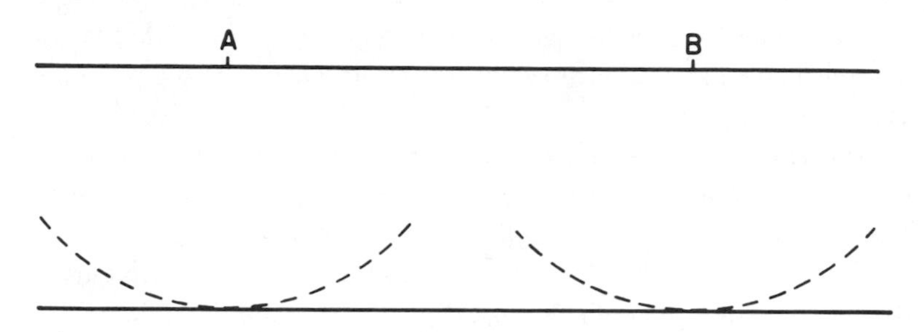

FIG. 33 Ruling parallel lines using arcs and tangent.

inches, on a rule, place the rule diagonally between lines *EAB* and *DCF* and mark off the eleven divisions along this diagonal line (*EF*). If side *AD* is parallel to the edge of the drawing board, lines parallel to *DC* and *AB* can be drawn through the marked points with a T-square.

Another method again uses diagonal lines but does not require a T-square. For example, a rectangle, *ABCD* (fig. 35), is to be divided into seven equal parts. Find seven convenient divisions on a rule and locate this distance (line *ab*) between *EF* and *GH*. Another line (*cd*) equal in length to line *ab*, also located between *EF* and *GH*, will be found to be parallel to *ab*. When the divisions previously plotted on the diagonals are connected, the connecting lines will be parallel.

PRODUCING SYMBOLS

Symbols such as crosses, stars, triangles, and circles may be used on a map or graph to plot different quantities. The variety of press-on and "dry transfer" symbols is enormous (fig. 36). These may be had in different sizes and colors, in black with no surround, and in black with a

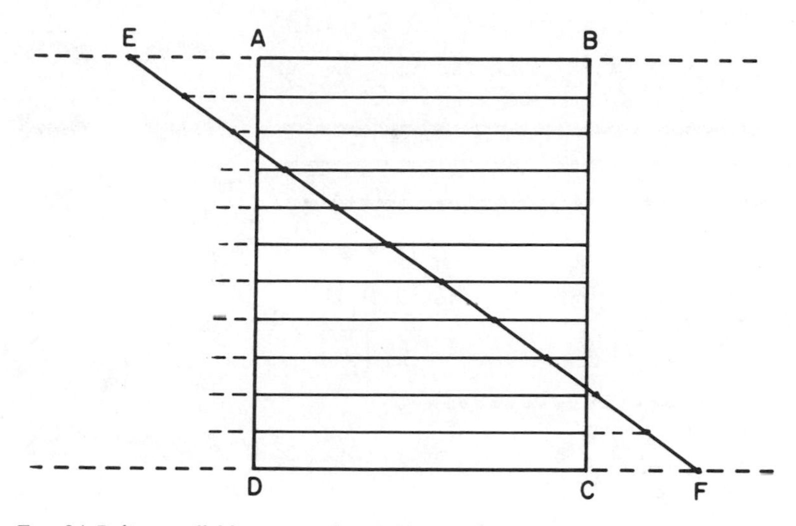

FIG. 34 Ruling parallel lines using diagonal line and T-square.

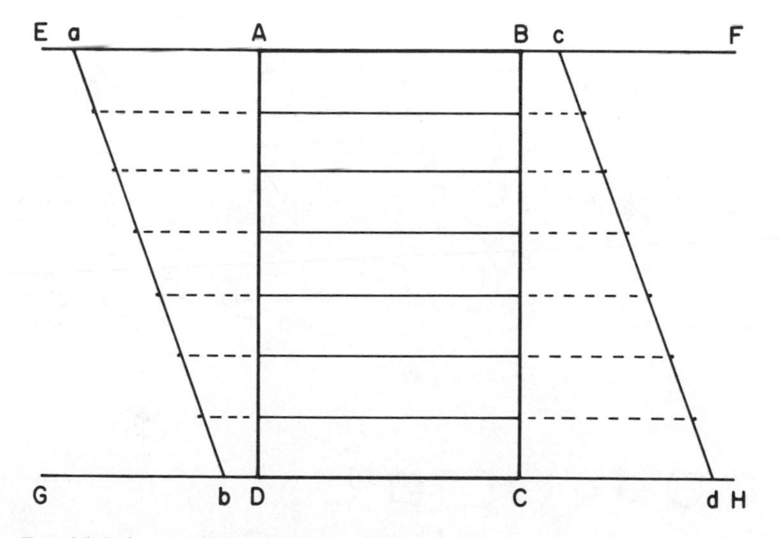

FIG. 35 Ruling parallel lines using two diagonal lines.

Fig. 36 Various press-on symbols and designs.

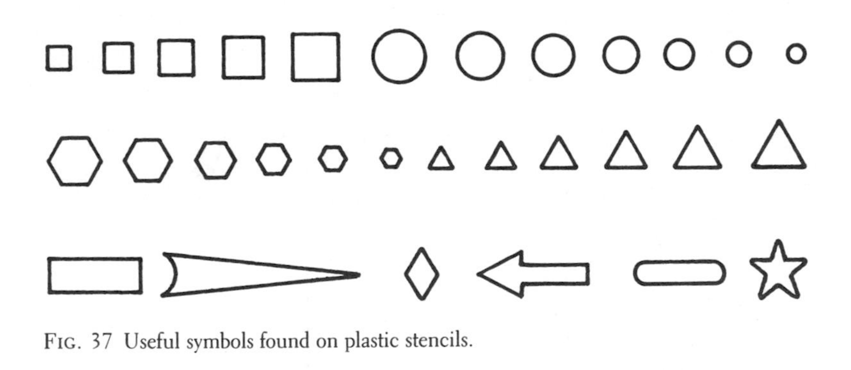

Fig. 37 Useful symbols found on plastic stencils.

Fig. 38 Some symbols found on the Leroy mapping template.

white background. In general, beware of old products, since these commonly break and flake and cause havoc. If in doubt, purchase fresh ones. (See p. 105 for directions for use and cautions about these labor-saving products.)

A sometimes more suitable solution (less expensive and more permanent) is to draw the necessary symbols with a template or stencil. Plastic stencils with appropriate symbols of various forms and sizes are readily available and, when used with a technical pen or Leroy socket-holder pen, make neat and uniform symbols (see fig. 37).

A standard scriber lettering set may include on its templates some useful figures such as circles and crosses, and special-purpose templates, such as for cartographers and engineers, contain a variety of useful symbols (fig. 38).

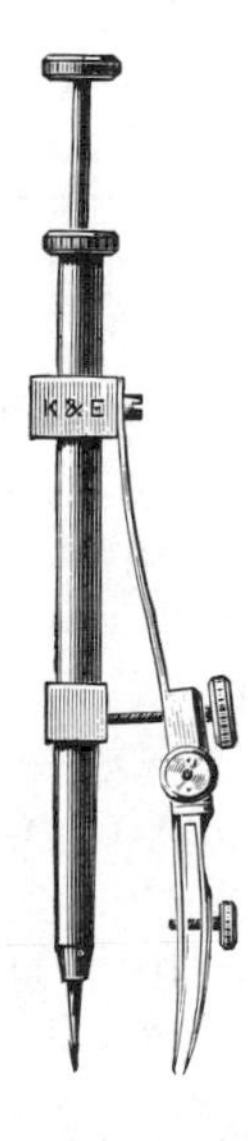

FIG. 39 Drop compass.
(Courtesy Keuffel & Esser Company)

Larger circles can be drawn with the aid of a drop compass (fig. 39), but practice before using this instrument, since unfamiliarity may cause blotting. Several companies make a compass for use with technical pens.

In general it is best not to let two symbols touch on a map or graph. Where two points to be plotted fall so close together that there is not enough room for two complete symbols, one should overlap the other.

Ideally, a clear space should separate the superior (complete) symbol from the inferior. This will help prevent the symbols from running together and becoming illegible when reproduced. Similarly, symbols should be dominant over lines and other markings on illustrations; that is, where a symbol and line come in contact, break the line so the symbol stands clear (see figs. 43, 44, 55).

Symbols on maps or graphs may need explanation but may not be readily available to the printer to set in a legend. For this reason the illustrator should explain such symbols in the illustration itself, usually in a short table printed below the map.

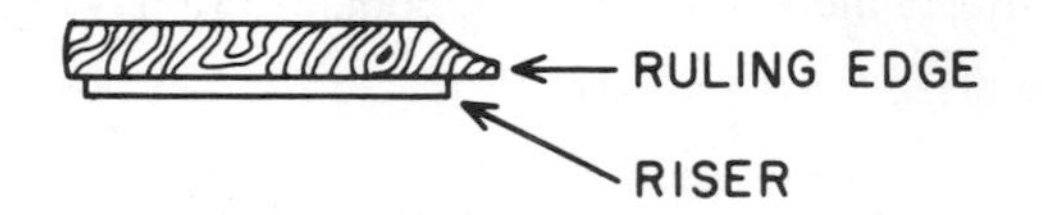

FIG. 40 End view of a rule raised from the drawing surface.

USE OF RULES AND TEMPLATES

A straightedge, curved rule, or template to be used with a pen must be raised from the drawing surface to prevent smearing the ink. If the rule is not beveled, a strip of blotting paper or thin cardboard can be glued to the underside with rubber cement, as shown in figure 40. This riser should be set back $\frac{1}{8}$ to $\frac{3}{16}$ inch from the ruling edge. On a template, glue several narrow strips of cardboard to the underside to avoid blocking the holes and yet provide a firm footing.

Remember not to slide the rule away from the wet line; lift it free of the paper when changing its position. A safer procedure is to leave the rule in place until the inked line is dry. Also, do not draw a second line to or from one that is still wet lest a blot occur at the point of junction.

SHADING PATTERNS

The least satisfactory method of producing a pattern over all or part of a graph or map is to draw it by hand. The result is most often an uneven, amateurish job.

Transparent overlay films are the most common and convenient way to provide areas of pattern on graphic materials. A multitude of overlays are available, printed in black, white, and colors, in lines and stipple, and in patterns of vegetation, rock, brick, waves, and so on (fig. 41).

Press a piece of film over the area to be shaded. Smooth it with finger pressure from the center to the edges, but do not burnish it down yet. With an X-Acto blade no. 11, carefully cut around the outline of the area. (Shading film may shrink very slightly—try to cut just on the outer edge of the drawn outline to allow for this.) Peel away the film outside the patterned area.

If there are letters or symbols within the shaded area, outline these with the X-Acto blade and remove the bits of film—lettering and symbols should appear in a clear, unpatterned space; see figures 54 and 55. An alternative (emergency) method is to do the letter on white paper, cut it out, and paste it in place on the completed illustration; this is not recommended for regular use, since pasted bits tend to fall off easily.

Caution: The adhesive used on some transparent overlay films is so "tacky" that it will take up any inked lines it touches. You may discover that when you remove the film from the symbols you drew so carefully you have removed the symbols as well. "Low tack" films are the safest; these include wax-backed films. Among satisfactory brands are Chartpak, Zip-a-Tone, and Formatt.

When all unwanted pattern film has been removed from the illustration, burnish the film in place with any smooth, blunt instrument or the back of your fingernail. If some tacky adhesive is left on the copy, gently wipe the surface with a bit of cotton moistened with rubber-cement thinner (a very flammable product).

Where two or more patterns are used in a single figure, try to separate those that are similar in value. A dot pattern and a line pattern may seem sufficiently different at first glance but may look very similar when reduced. When choosing patterns, squint and try to find a difference in value.

In general, line patterns reduce better. Another advantage to using a line pattern is that by changing the angle and by placing a second layer of pattern over a first, one line pattern can be used to differentiate several areas (fig. 42). If the film is too old or too "low tack," the layers may not adhere to one another; check this before you send off the illustration.

TRANSFERRING

In the previous chapter I discussed carbon and tracing-table methods of transferring a sketch to the final ground (see p. 40). Both methods are effective in preparing maps and graphs. If you draw the map or graph on tracing vellum or frosted film, however, there will be no need to transfer at all.

Graphs

The purpose of a graph is to present a comprehensive and lucid picture of data and of relationships between different data sets. A given set of

FIG. 41A Various shading-film patterns, original size.

data can often be presented graphically in several ways. So the illustrator will be familiar with alternative methods of graphing data of various sorts and thus can choose the most appropriate illustration, the more important kinds of graphic presentation are illustrated and discussed below.

SCATTER DIAGRAMS

In the most common type of scatter diagram, a point represents two quantities whose values are scaled on perpendicular coordinates. The horizontal coordinate is known as the *x*-axis, or abscissa, and the vertical is the *y*-axis, or ordinate.

Customarily, the lowest values are at the left of the *x*-axis and at the

FIG. 41B The same shading-film patterns as shown in figure 41A, reduced one-third off.

bottom of the *y*-axis. Usually it is necessary only to number the values along the bottom for the *x*-axis and up the left side for the *y*-axis, though clarity may occasionally require that the values be shown on all sides. The explanatory label along the left *y*-axis should be placed so that it reads properly when the graph is rotated 90° clockwise.

The labels along the coordinates are usually printed all in uppercase letters, except for abbreviations such as hr or min. Uppercase lettering reduces clearly.

Important: Plan all lettering so that no letter or number will measure less than 1.5 millimeters in height when reduced for publication.

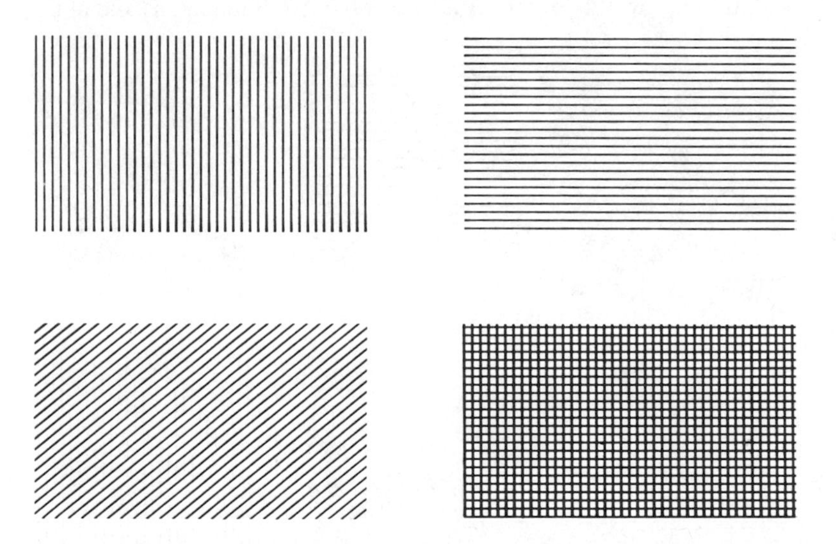

FIG. 42 Single film pattern used four ways.

There is no set rule on whether grid-coordinate lines should appear on the face of the scatter diagram. Usually a glance at the values scaled in the margins of the graph will allow the reader to estimate the quantities indicated by any particular point on the graph. Where more precision in reading is desired, some basic intervals may be ruled (see "Ruling Parallel Lines," p. 71). If great precision is called for, it will be best to draw the graph on graph paper printed in black or red, which show up in reproduction.

The example in figure 43 is a scatter diagram in which internarial distance of several individuals is plotted against snout-vent length. Individuals of one or more categories may be added to the diagram and plotted with a different symbol, as in figure 47, where female and juvenile specimens have been graphed along with the males represented in figure 43. The variety of categories that could be included in a graph is restricted only by the number of symbols that may be devised. Clarity, however, demands that the number be limited—two easily read graphs are better than one that is too complicated.

The graph discussed above is based on a system of perpendicular co-

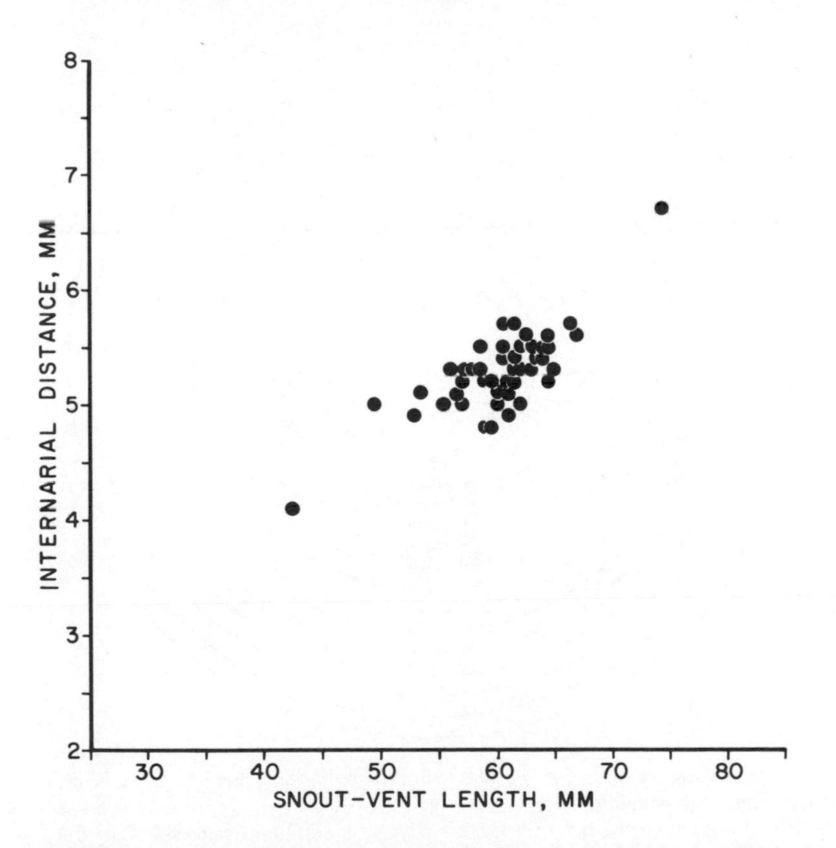

FIG. 43 Simple scatter diagram, with several individuals of a single class plotted. Measurements are of the frog *Nyctimystes disrupta*.

ordinates. It is also possible to construct a scatter diagram on angular and polar coordinates, in which one value is determined by the direction the radius takes and the other value is scaled along the radius. This sort of diagram is shown in figure 44.

LINE DIAGRAMS

Where data are graphed as frequency of occurrence on the y-axis of some variable scaled on the x-axis, it makes for a clearer presentation if

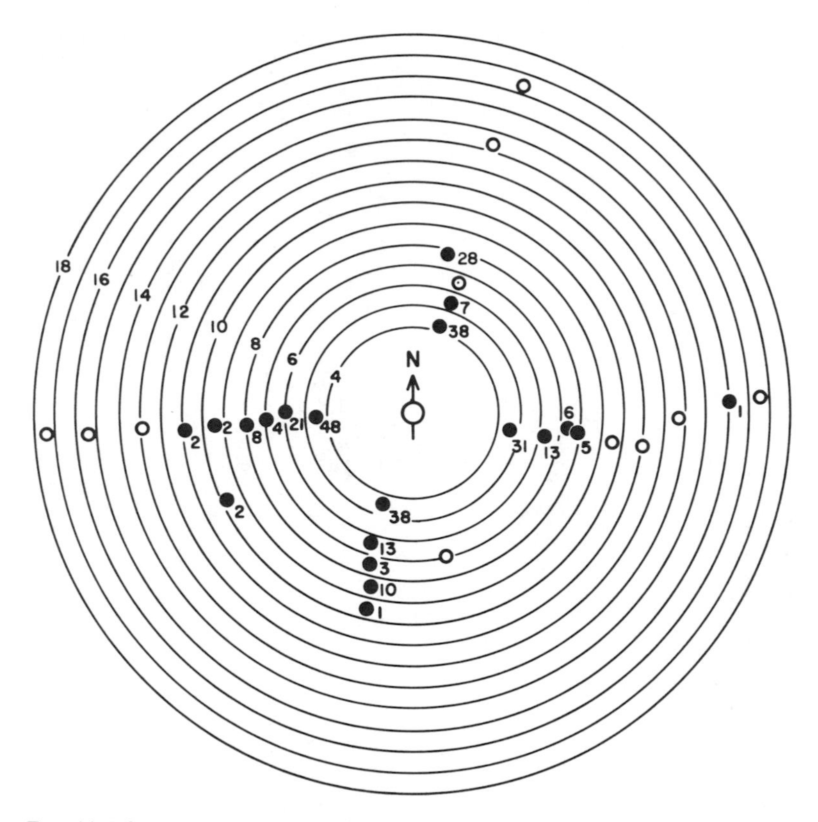

FIG. 44 Polar scatter diagram. The plots indicate traps in which flies released at the central point were recaptured. The number of flies is shown beside each solid dot; open circles mean an empty trap. Concentric circles represent 2-mile intervals. (Data from F. C. Bishopp and E. W. Laake, *Journal of Agricultural Research* 21, no. 10 [1921]: 729–66)

the points are connected to form a continuous line. Since the line forms one or more irregular figures, this graph is known as a frequency polygon. An example is shown in figure 45.

When raw data are used, very irregular polygons may result. It is proper to group data, which often will make smoother polygons. In the example in figure 46, the data of figure 45 have been replotted and grouped in intervals of 3 millimeters, although the original data were recorded in intervals of 1 millimeter. Note also that the *y*-axis has been changed from an absolute to a relative scale. The choice of absolute or relative (percentage) numbers for the *y*-axis depends on circumstances involving presentation in the accompanying text and comparison with other graphs.

Often the data presented in a scatter diagram may be expressed in the form of a line that is estimated by eye or calculated mathematically to average the many point observations. Figure 47 shows lines that by cal-

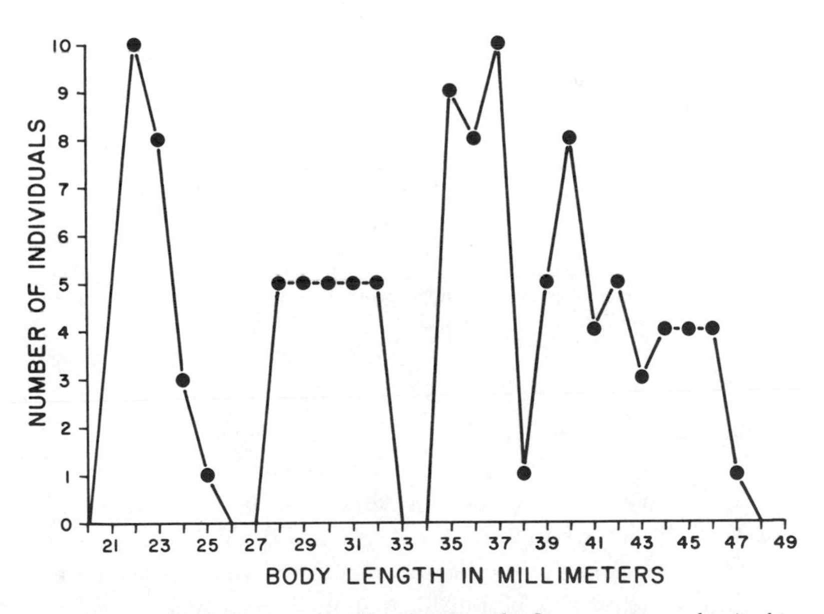

FIG. 45 Frequency polygon, a form of line diagram. This figure presents raw data (in this case the snout-vent length measurements of 109 lizards, *Xantusia vigilis*) without grouping or other modification. See figures 46 and 48.

culation illustrate the average trends for the data presented in scatter form in figure 43. As was done here, a good practice is to combine in one graph the individual points of the scatter diagram with the lines representing those points.

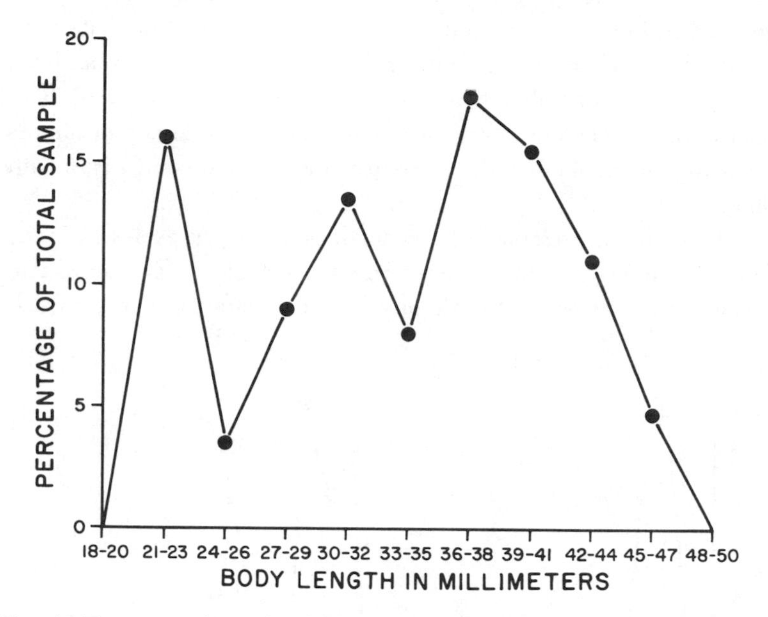
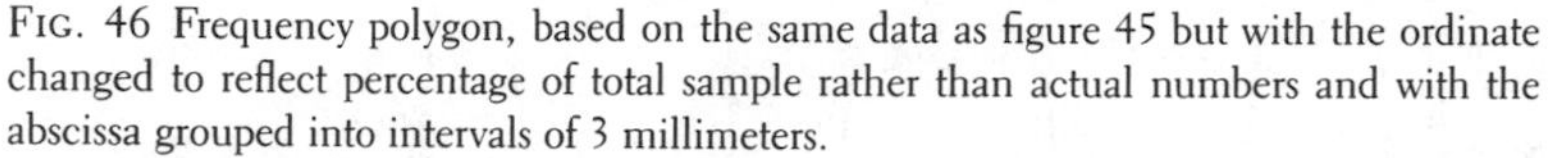

FIG. 46 Frequency polygon, based on the same data as figure 45 but with the ordinate changed to reflect percentage of total sample rather than actual numbers and with the abscissa grouped into intervals of 3 millimeters.

BAR DIAGRAMS

Bar diagrams are graphs in which a line or rectangle represents each category. A histogram is a variety of bar diagram that represents frequency data in a form that is often more easily comprehended than the frequency polygon. Figure 48 is a histogram based on the same data as the frequency polygon seen in figures 45 and 46. An additional refinement is that the number of individual animals in each class has been added above each bar of the histogram. The rectangles can touch, or they may be slightly separated by a narrow white space, as in the figure, depending on the taste of the artist and the area available for illustration.

In a much-used and adaptable form of bar diagram, a line represents the range of values of a series of observations, and appendages to that line illustrate other statistical data pertinent to the sample (fig. 49).

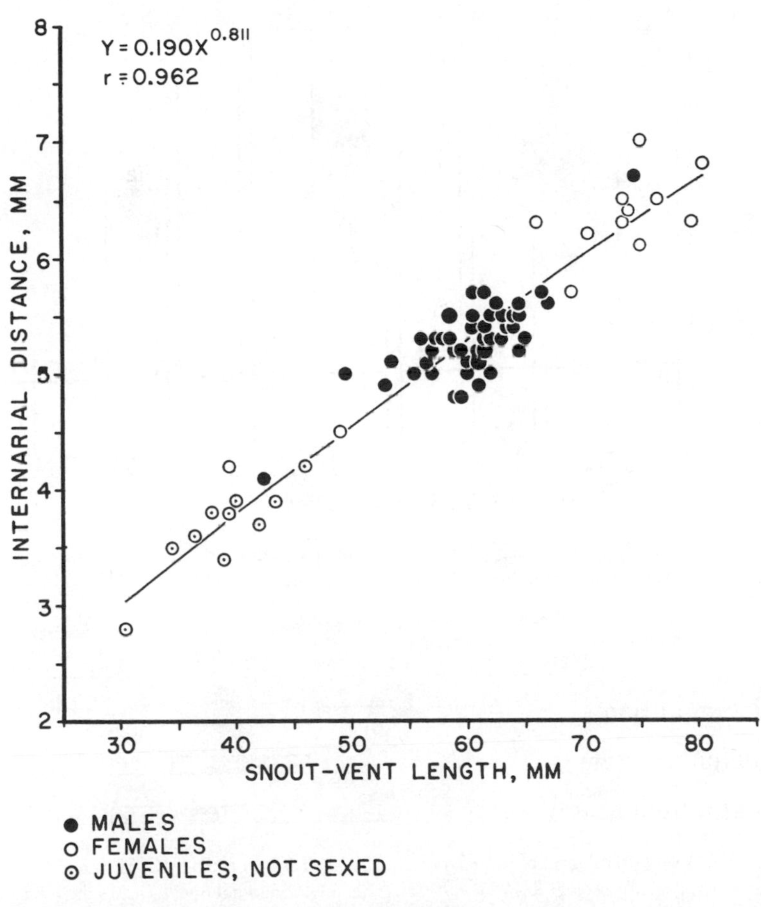

FIG. 47 Combination of line and scatter diagram techniques. Two more classes (females and juveniles) are added to the data of figure 43, a mathematically calculated line of best fit shows the average trend of the relationship, and the regression formula and correlation coefficient are given. (From R. Zweifel, *American Museum Novitates*, no. 2759 [1983]: 1–21)

Commonly, a bar or crossbar at a right angle to the line represents the arithmetic mean, and rectangles on each side of the bar represent such statistical quantities as the standard error of the mean and the standard

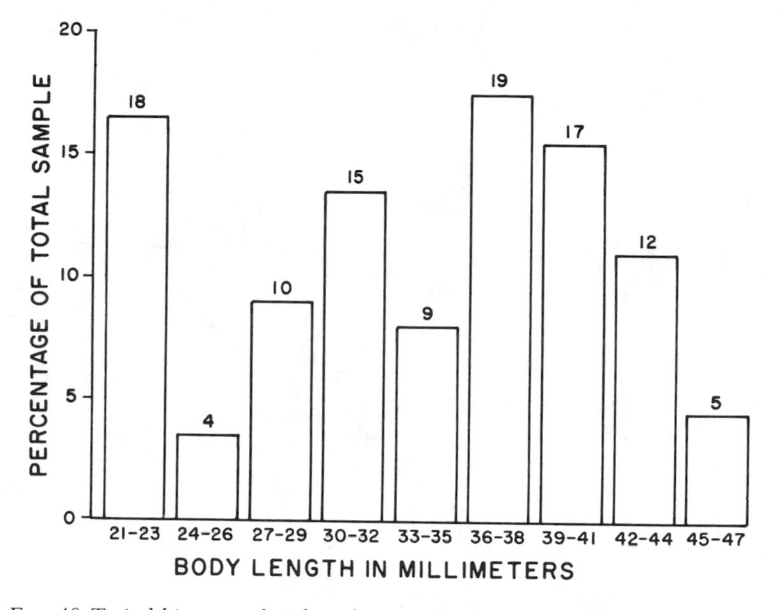

FIG. 48 Typical histogram, based on the same data as figures 45 and 46 and with the same axes. The number of individuals in each 3-millimeter group is shown.

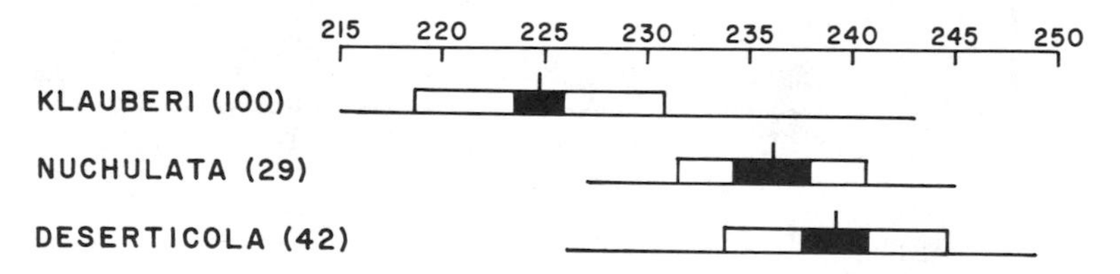

FIG. 49 A very useful form of bar diagram, constructed as advocated by C. Hubbs and C. Hubbs (*Systematic Zoology* 2 [1953]: 49–56). The long horizontal line indicates the range of variation in a given sample; the vertical line indicates the mean. In this example the statistical quantities of a standard deviation and two times the standard error on either side of the mean have been indicated by the light and dark rectangles, respectively. The graph is based on unpublished data for the number of ventral scales in three subspecies of the night snake, *Hypsiglena torquata*.

deviation or some multiple of these statistics. The common practice of making the diagrams symmetrical, with mirror-image boxes on both sides of the range line, conveys no more information and wastes space.

This type of diagram is particularly useful when several samples are compared. In illustrating geographic variation, it can be effective to combine this form of bar diagram with a map showing the geographic origin of the several samples (see fig. 50).

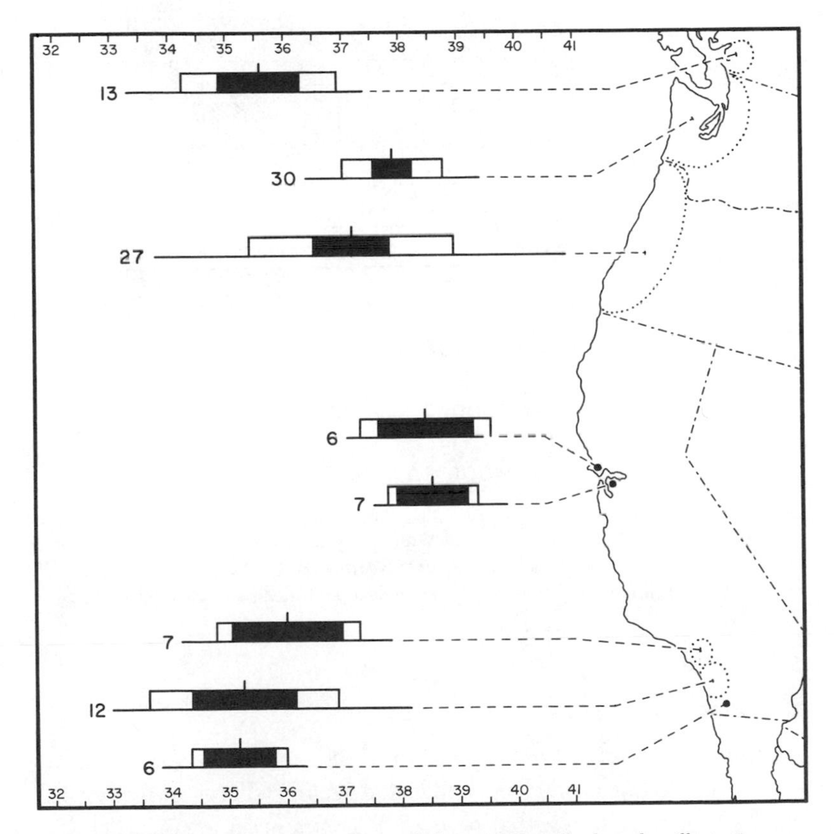

FIG. 50 The form of bar diagram shown in figure 49 is well adapted to illustrating geographic variation where the samples are arranged in a more-or-less linear fashion. (Data are from R. G. Van Gelder, *Bulletin of the American Museum of Natural History* 117, art. 5 [1959]: 229–392)

Bar diagrams may be used in a variety of ways to present and compare data. Figure 51 shows the relative importance of three principal prey species in the diets of several predators. The use of different shading patterns to represent the prey species makes it immediately evident who eats how much of what. Also, it is possible to compare the food habits of the various predators, all of which live in the same region. Again, the illustrator may prefer to separate the side-by-side rectangles by white space.

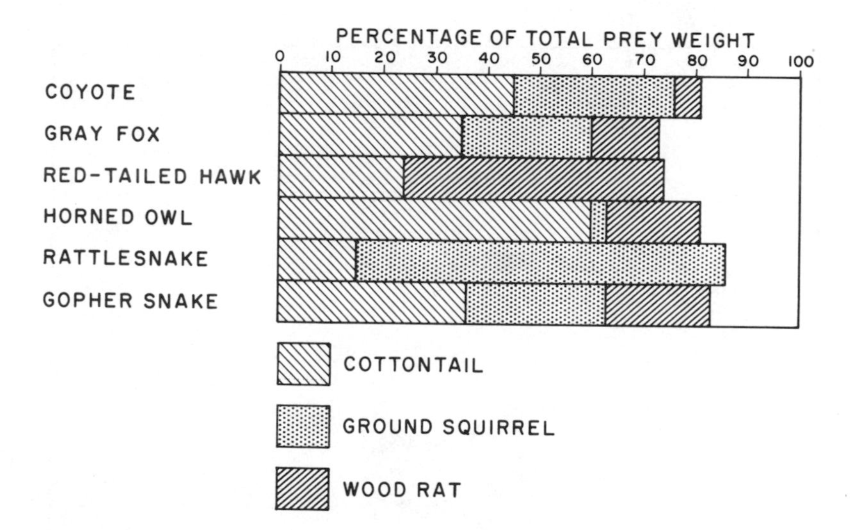

FIG. 51 The use of shading patterns to differentiate segments in a bar diagram. The contrasting patterns make it easy to compare the relative importance of a given item of prey in the diet of several species of predators. (Based on data published by H. S. Fitch)

AREA DIAGRAMS

The most commonly used area diagram is the pie or sector diagram, in which a circle (pie) is divided into radial sectors (slices), as shown in figure 52. The data presented here are the same used to construct the comparable part of the bar diagram in figure 51.

In drawing a pie diagram, extend the first radius vertically upward from the center. If there are to be many radii converging at the center, cut a small circle out of the center of the pie to avoid congestion (and possible blotting of ink). A circular protractor is handy when drawing pie diagrams.

PICTORIAL DIAGRAMS

Pictorial diagrams use pictures in a variety of ways, often combined with other types of graphs. For example, in a pie diagram, a drawing of the real object might be enclosed in each sector to make its meaning plain.

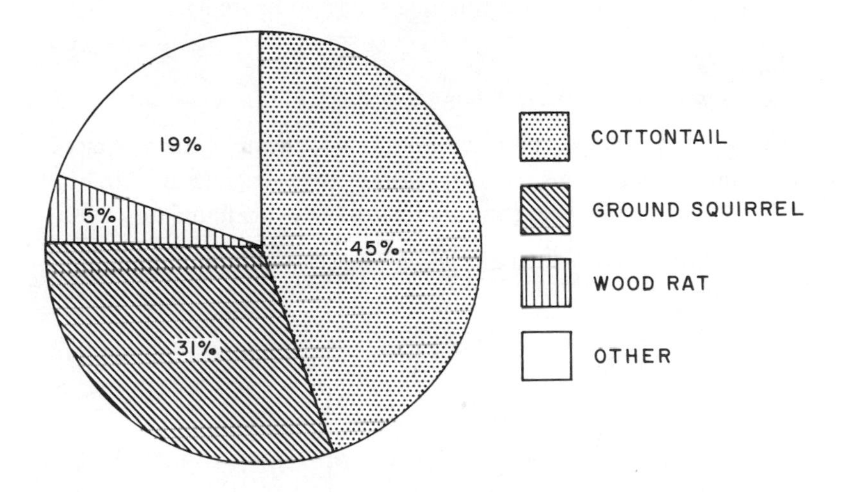

FIG. 52 Pie (sector) diagram illustrating the percentage of total prey weight of three important species in the diet of the coyote. These are the same data presented in figure 51 and point up the superiority of the bar diagram where it is necessary to compare sets of data.

Often a standardized picture is altered in size to indicate quantity. This may be done in toto, the figure simply being magnified or reduced, or a standard-sized figure may be whittled away piece by piece. Doubling the linear dimensions of a figure, of course, will increase the area of that figure much more than twice, a consideration commonly disregarded in advertising art.

Pictorial diagrams find most use in popular or semipopular publications, where things are made as easy as possible for the reader. Such diagrams often lack precision and may be superfluous in scientific publications. Effective use of pictorial diagrams has been made in research publications, however, especially in illustrating phylogenetic charts and identification keys.

THREE-DIMENSIONAL GRAPHS

Some types of graphic presentation are adaptable to the addition of another dimension to the x- and y-axes common to so many graphs. An example of this sort of modification is shown in figure 53.

Maps

USES OF MAPS IN SCIENTIFIC PUBLICATIONS

One of the most important uses of maps is to illustrate the geographic distribution of organisms. This may be done in a general way by indicating with shading film where the animals or plants are found (fig. 54). As

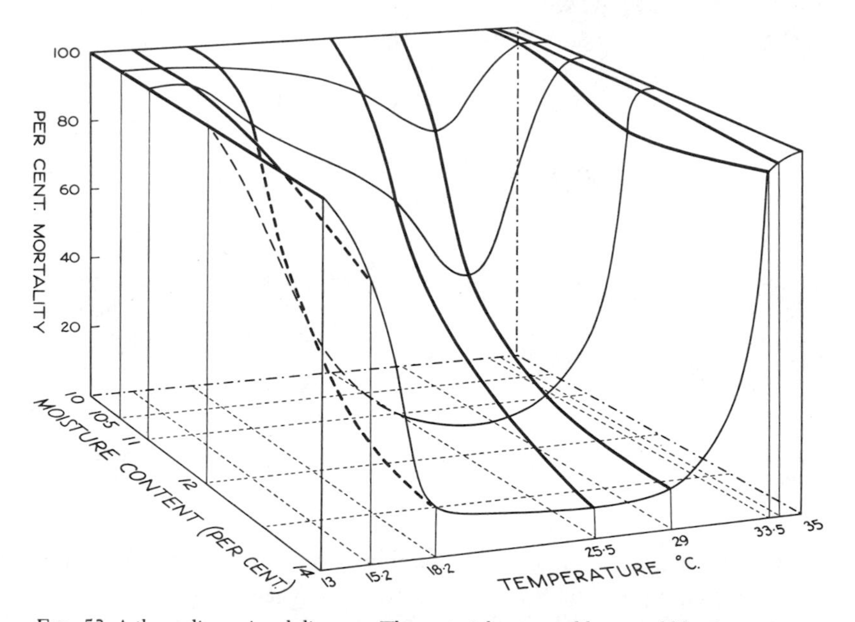

FIG. 53 A three-dimensional diagram. The material presented here could be shown in a series of five diagrams of temperature against percentage of mortality (one for each level of moisture content), but the correlations among the three variables would be less easy to visualize. (From H. G. Andrewartha and L. C. Birch, *The Distribution and Abundance of Animals* [Chicago: University of Chicago Press, 1954])

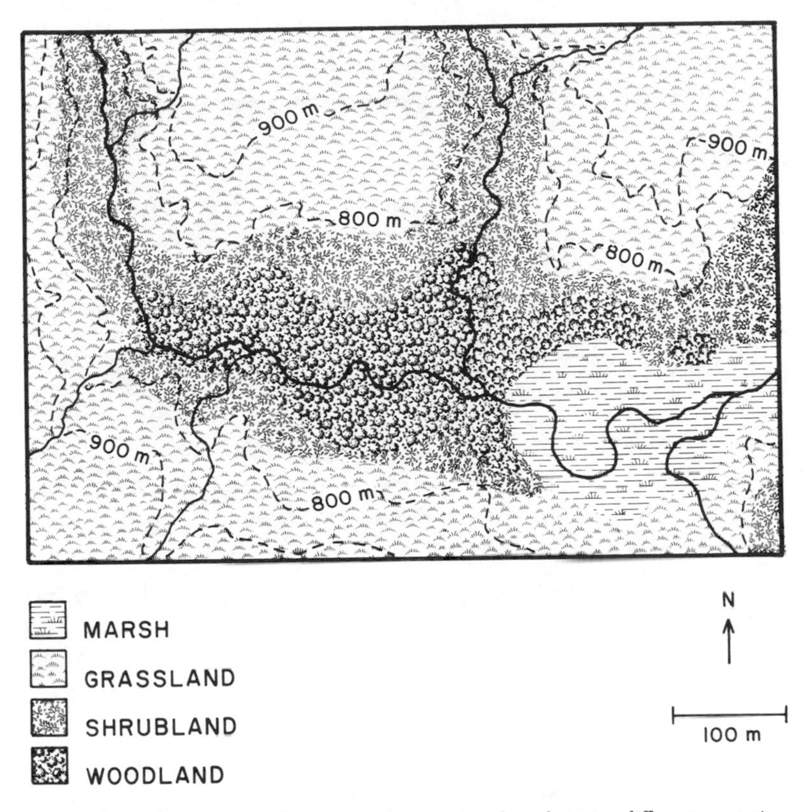

FIG. 54 The use of shading-film patterns in mapping, here featuring different vegetation types.

an alternative, the area where the organisms are not found may be shaded and the area of occurrence left clear.

The author must delimit the area to be shaded, of course, but the illustrator must use good judgment in choosing the shading pattern. It must be bold enough to be distinct from other detail on the map, and you must keep in mind the reduction the map will undergo in reproduction. Often several patterns will be required on a single map, and these must be easily distinguished when reduced. If the illustrator's budget is small, it is well to keep in mind that one carefully chosen line pattern can be used in several ways on the same map (see fig. 42).

On a spot map, individual localities are shown by one or more symbols. If more than one kind is to be used, choose symbols sufficiently different that they are readily distinguished when reduced.

An effective form of distribution map combines symbols and shading patterns. When you draw this sort of map, the shading pattern on and immediately surrounding each symbol, and around any printing on the map, must be removed so they will not be obscured. (See p. 77 for precautions when selecting adhesive pattern films.)

In mapping vegetation types and other habitat classifications, you will find a wide choice of shading film patterns. Often you can find a pictorial pattern, such as for woodland, marsh, grassland, or water (see figs. 41, 54).

SELECTING A BASE MAP

The simplest form of map is an outline that shows nothing more than major physiographic boundaries or political subdivisions. Such a map is often used as a base for a biological illustration. At the outset, be advised that if you intend to use a commercially published outline map as a base, you should take care to obtain written permission from its publisher before allowing your work to appear in print, to avoid copyright infringement.

Most often the artist who must prepare a map for a scientific publication will find it necessary to make a new base map. This may be done by using enlarging, reducing, and tracing, as discussed in earlier chapters, on commercially available maps.

The base map upon which data are plotted must be selected with care. The first consideration is to eliminate superfluous detail. A base map may be nothing more than an outline, or it may be complex, showing topography (by contours and hachure), vegetation, and human cultural features of the landscape. Obviously it is best to include only details that serve a useful purpose. Elaborate detail of drainage systems might be appropriate for a map showing the distribution of aquatic organisms but would be out of place on a map dealing with terrestrial animals. In the latter instance, a few large river systems (for orientation) might be sufficient.

The base map should show no more area than necessary; it would be a waste of space to show the entire United States in mapping the occurrence of a species confined to the West Coast. If it is necessary to relate the West Coast region to the whole of the continent, this may be done in a small inset, as discussed below.

GEOGRAPHIC ORIENTATION

If the area mapped is not readily placed in its broader geographic context, include a small inset that shows the main map in relation to a larger area. For example, a restricted area in Arizona is more easily visualized if you include an inset base map of the entire state (fig. 55, *right*).

On any map, include some features that will allow geographic orientation, even when these features in themselves are not significant to the basic purpose of the map. Cities, major political boundaries, or rivers will locate the mapped area in relation to the rest of the world. Lines of latitude and longitude may be indicated around the borders; if these do not conflict with the material in the body of the map, they may be drawn across the face.

COMPASS DIRECTION

It is a convention to orient maps with north at the upper edge. Occasionally available space and the shape of the area mapped may dictate some other orientation. In either event, the illustrator should include a marker to indicate north. This need not be elaborate; a simple arrow with N printed below it is sufficient.

SCALE

Every map, whether the area depicted is measured in square feet or thousands of square miles, should specify the scale. Scale may best be shown graphically by a line or bar subdivided into segments proportional to the distance represented on the map; this is drawn on the face of the map (see fig. 55). This graphic method has two distinct advantages over other methods of indicating scale: the scale undergoes reduction with the map when it is printed and so remains in proper proportion; and the map user will not be required to measure or calculate further.

Scale may also be given in words in the legend; for example, "2 inches: 1 mile." The colon means "is proportional to" and is preferred to an equals sign.

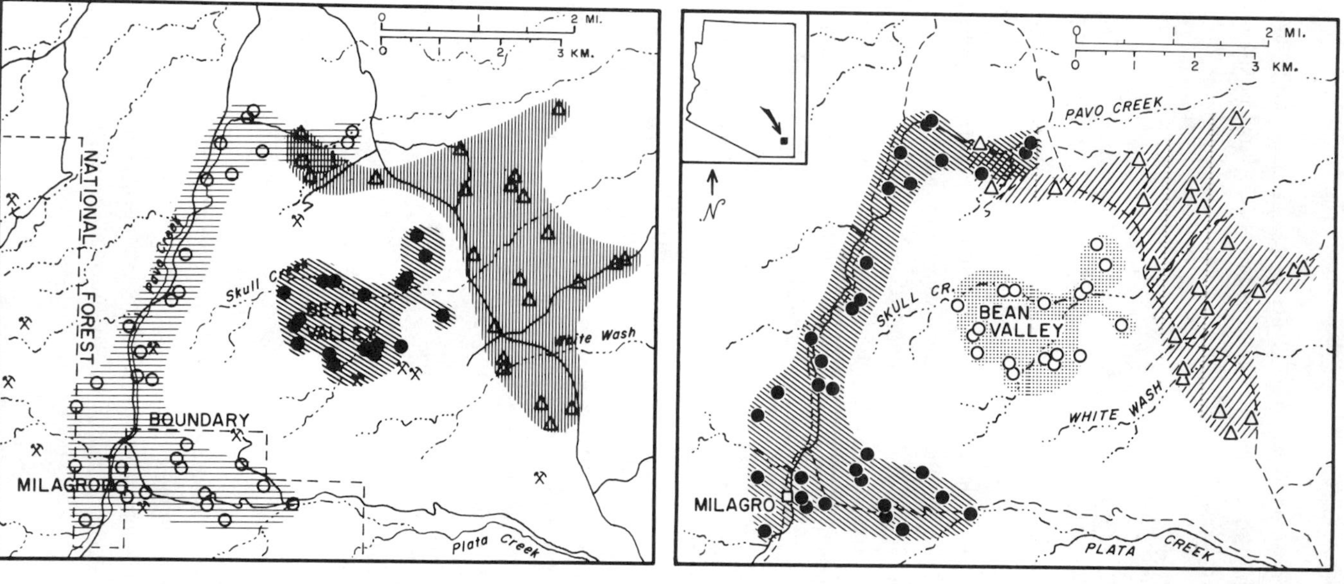

FIG. 55 Some common errors in mapping. In the map on the left the numbers in the scale are too small to stand reduction, letters and numbers are obscured by shading patterns, and symbols run together and overlap. The correctly drawn map on the right omits extraneous detail, differentiates between creek and road, and locates the map area on an inset of a wider geographic area.

A representative fraction such as 1/5,000 may also indicate scale, in this instance meaning that one unit on the map is equivalent to five thousand units in the field (the map is one-five-thousandth the size of the real area).

LETTERING

The mechanical aspects of lettering, as well as cautions and precautions, are treated in chapter 7. Here I will mention some peculiarities of wording on maps.

Whatever the style or capitalization, no letters or numbers should reduce to less than 1.5 millimeters when the map is reproduced for publication. Plan the map so that the smallest numbers and lettering, such as on the scale, will adhere to this restriction.

It is easier to read a map when the words on it are arranged parallel with its lower border. At times, however, lettering should follow some natural physiographic feature such as a river or a mountain range. Lettering on a curved line is easier to do with press-on or dry-transfer letters than with the Leroy system.

In general, uppercase lettering is used for the most important words; for words next in importance, the initial letters should be capitals. Since capital letters usually reduce better than lowercase, you could show relative importance by using all capitals in different type sizes, as long as these do not reduce to less than 1.5 millimeters for publication.

The style of lettering chosen for a map should be open and clear, easy to read quickly. The Leroy sans serif style fits this description; so does Helvetica Light, available in many dry-transfer and press-on brands.

Rivers and other bodies of water are conventionally labeled in italics. Most lettering scribers will adjust to this modification.

When drawing a series of maps for the same publication, there should be one standard style throughout. This includes the choice of line widths for borders, political area outlines, rivers, pattern area outline (if there is one), and so on. In addition, political boundaries should conform to a uniform pattern (dot-dash for state lines, for instance, and dot-dot-dash for international boundaries); use wider lines for some designations than for others. Keep in mind the reduction each map will undergo and adjust the sizes and widths accordingly.

When a word on the face of the map is associated with a symbol, the lettering should be set off slightly to the right (or left) and below the symbol to avoid confusion of letters and symbol (see fig. 55). Remember that letters and symbols take precedence over other material on the map, and that lines should be broken where they conflict with them.

BORDERS

A single or double rectangular border is advisable for any map. Where degrees of latitude and longitude are included, the numbers may be placed within a double border; otherwise a single border is sufficient. Without a border of some kind, a map tends to look unfinished. An added note of elegance is to draw the right and lower borders very slightly wider than the top and left borders; this simple touch will "lift" the map from the page in a three-dimensional effect (see figs. 4, 56).

COMBINING GRAPHS WITH MAPS

The use of graphs in conjunction with maps has been mentioned (see p. 87). The form of graph shown in figure 50 is best adapted to illustrating geographic variation where there is a more-or-less linear arrangement of localities. Sector and bar diagrams may also be adapted for use with maps.

OVERLAYS

Sometimes it is necessary to publish a map or other illustration in two or more colors; for example, a map featuring drainage might emphasize the rivers and lakes in blue while the remainder of the map is in black and white.

For publication, everything is drawn in black; each color must be drawn separately. First draw the black portion of the map (the base map) on any chosen ground. On a transparent ground (tracing vellum, frosted polyester film) taped in position over the base map, now draw the blue portion of the map, *also in black ink*. This includes pattern areas, arrows, lettering, and every line or dot that will be printed in blue on the published map.

Both base (to be printed in black) and overlay (to be printed in blue) have register marks placed in the corners outside the field of the figure (see fig. 56). These must be placed precisely, so that when the overlay is in correct position on the base, the register marks coincide exactly. The printed map must be run through the press once for the black base and an additional time for each overlay color. The register marks are the guides that allow the pressman to ensure accurate register in the two or more printings.

If there is no type on a map, it is easy for the printer to print it upside down or backward ("flopped"). To prevent such an error, rotate one of the register marks differently from the others or place the marks on the artwork so they are not parallel. Also, write "Top" at the top of the illustration, outside the area to appear in print.

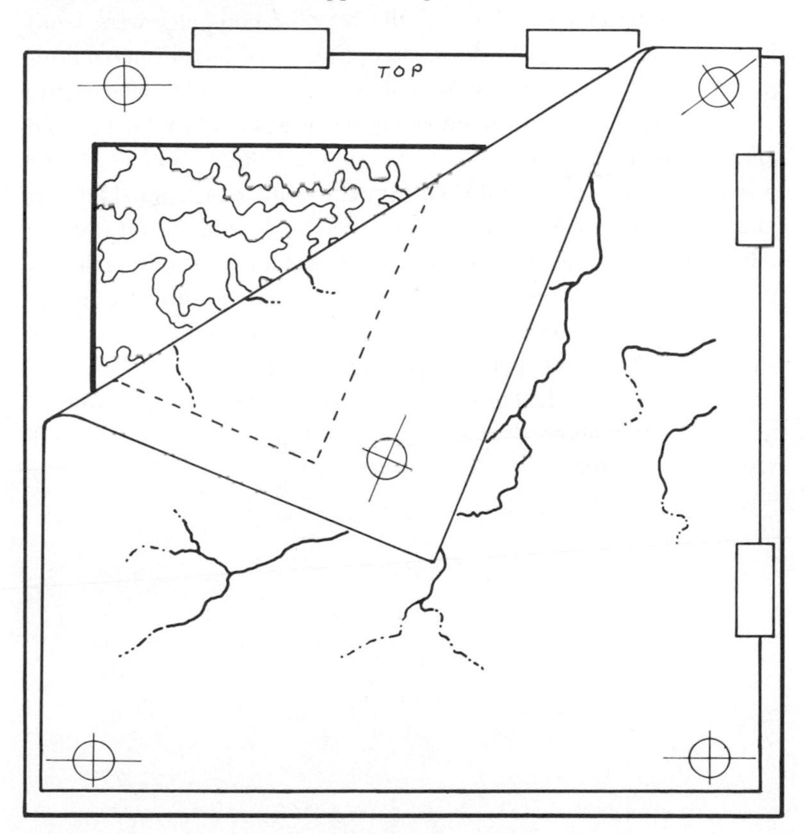

FIG. 56 Making an overlay, using register marks to ensure accuracy.

Sometimes you can draw the overlay in color, if that color will photograph as black. Certain red inks can be used; there are overlay films printed in red, in solid sheets (such as Rubylith), shading patterns, symbols, and lettering. Two advantages to doing the overlay in color are that you can see through to the black base while working and can judge the final effect.

Computer-Generated Graphics

Personal computers and software that can generate graphs are readily available to many illustrators. Instruction in the use of such equipment is beyond the scope of this work, but a few comments are appropriate.

At one time graphics done on the personal computer were reproduced in hard copy only by dot-matrix printing. Although such a printout is functional, it is far from elegant, and many editors properly refuse to accept it as a substitute for conventional artwork. With the advent of computer driven pen plotters and, later, laser printers, graphics can now be drawn that in no respect must defer to those done by hand. The illustrator who is fortunate enough to have these tools at hand and is skilled in their use still must pay close attention to the fundamentals of designing graphs for publication; mechanization does not substitute for artistry.

The foregoing negative comments notwithstanding, dot-matrix graphics can be useful. For example, such a graph may be useful as a base for tracing a finished product in ink. Remember that you can use a variable-magnification copying machine to adjust the size of the crude copy before you trace it.

7

Lettering

Introductory Remarks

Many biological illustrations require lettering. Parts of drawings and points of interest must be labeled; graphs and maps need numbers, place-names, legends, and titles. Some tables have to be lettered completely, though these are more usually set in type. It is no exaggeration to say that the quality of lettering on an illustration plays a major role in its general appearance even though the lettering may represent a minor part in terms of area covered and effort expended. The illustrator, then, should pay particular attention to producing neat, attractive lettering.

Some editors prefer that the artist do no lettering at all on any illustration, using instead an overlay sheet to show the position and wording of legends, labels, and directions of all kinds, which will be set in type by the printer. The overlay sheet may be tracing vellum or acetate or polyester film. The artist must place register marks outside the margins of the illustration itself and put corresponding register marks on the overlay sheet. Unless the register marks match exactly, the labels will be printed out of position on the illustration (see p. 96 for making an overlay).

In place of an overlay, some editors ask the artist to send, along with the illustration itself, a photocopy of the illustration bearing all necessary numbers and wording in proper position.

SIZE OF LETTERING

Lettering should not be either so large as to crowd the field or so small as to appear lost in a vast blank area. Different sizes of lettering, and upper- and lowercase, can be used to categorize subjects in an illustration; that is, use larger letters and capitals for more important labels and information.

Keep reduction in mind when planning the size of all lettering. When reduced for publication, no letters or numbers should be less than 1.5 millimeters high. The smallest lettering on the illustration, then, may dictate the size of the rest.

If an illustration is to be photographed for a projection slide, see chapter 8 for determining size of lettering.

LETTERING ON THE DRAWING AND PASTED LETTERING

Lettering may be done directly on the copy, or it may be done separately and later fixed in place. Each method has its advantages and disadvantages. In general it is easier to do the lettering separately and then attach it in the appropriate positions on the map or graph. This eliminates the possibility of spilling or smearing ink on the artwork, centering words and phrases is simpler, and any mistakes can be corrected more easily than on the illustration itself. But pasted-on lettering is easily detached and lost, so some editors refuse such artwork. Another disadvantage is that shadow lines from the labels may appear when the work is photographed for publication, requiring extra work for the printer and added cost for the publisher.

Some of the problems of pasted-on lettering may be overcome if the lettering is done on Scotch 811 Magic Plus transparent removable tape (do *not* use regular Scotch Magic tape—the ink will flake off as soon as you try to pick up the tape). "Removable" tape can be lettered in any length line, then cut and pressed into position on the illustration. (If the ink "beads up" as you letter, gently rub a little pounce, or draftsman's pumice, on the tape and wipe it off before you proceed; this may reduce the slickness enough.) This tape is "low tack," so it can be picked up and moved without damaging a paper ground and without removing the inked lines or symbols on the illustration. (Even the lowest of low-tack adhesives, however, will pick up and destroy dry-transfer lettering.) Because the tape is tacky, be aware that it will pick up any dust, eraser crumbs, or tiny hairs in the vicinity; be sure to put it down only on an absolutely clean surface when you letter on it. Fingerprints may be picked up as well; leave enough space at the ends of words or phrases so you can cut away any smudged tape after you apply it to the final copy. In addition, be sure the inked lines on the illustration are thoroughly dry before applying any "removable" product over them, or they may peel off with the product.

White artist's tape can be used in this same manner, but its adhesive is more "tacky." Again, the tape may have to be "pounced" before lettering.

If you do decide to use pasted-on lettering, in spite of possible label shadows, there are several methods worth considering. An electric typewriter with a new, dense black carbon ribbon produces acceptable copy if the size of the type is appropriate for your illustration. High-quality dot-matrix computer printout can produce acceptable copy. To make it darker, do the printing larger than needed, then reduce it photographically to the desired size, thus sharpening and darkening the print. If you have access to a laser printer, many styles of print are available for use with a word processor. The Kroy lettering system impresses letters on a transparent or opaque adhesive strip. Letters, words, and sentences are cut from the strip and pressed in place on the artwork. Try out the "tackiness" before applying this tape to your copy. Many type styles and sizes are available, and the finished product may not be distinguishable from professional typesetting. Such a device is likely to be found only in large institutions with a substantial need for professional-quality lettering.

Hand Lettering

If the artist is skilled in lettering by hand—or better yet in calligraphy—this can be used to tasteful advantage on maps and some other scientific illustrations. Lacking such skill, you would do far better to use a more mechanical method of lettering like those discussed below.

Lettering Guides

Various stencils and templates are available, cut with letters, numbers, and symbols of all kinds and sizes (fig. 57). Koh-I-Noor and Wrico are two satisfactory brands; others will be found in art-supply stores. Technical pens, or the Leroy socket-holder pen, are used with the stencils and templates.

To place and space the lettering most easily, the stencil should be

FIG. 57 Plastic lettering stencil. (Courtesy Koh-I-Noor Company)

transparent plastic and used on a transparent ground (vellum or film) taped over grid paper. This procedure is slow and painstaking (you must wait for the ink to dry before doing the next letter); if you need much lettering, I suggest a different method.

Mechanical (Scriber) Lettering Systems

The chief advantage of using a mechanical, or scriber, lettering device is that the lettering will be uniform, and a person unskilled in freehand lettering can produce neat, attractive work with only a little practice.

The most evident disadvantages are that lettering may go somewhat more slowly than for an illustrator adept at freehand lettering, and the initial cost of a scriber lettering system is more than someone who letters only occasionally may wish to invest.

The deservedly popular Leroy mechanical lettering device employs templates and a scriber (fig. 11); the Koh-I-Noor company also makes a scriber system. One arm of the scriber is tipped with a metal point that fits into the letters and numbers depressed into the template; a second arm, also with a metal point, fits into a guide groove in the template; the third arm of the scriber holds a technical pen, an inkwell point, or a graphite point. By moving the first scriber point within the letters in the template and moving the template along a stationary guide bar, you can letter in a straight line with comparative ease.

The simplest scribers can make only vertical letters. A scriber with an adjustable arm can produce slanted letters (italics), and more complex scribers can make tall or short, vertical or sloping (forward or backward) letters with a single template.

Templates are available with letters of different sizes, with numbers, and with various symbols. Capitals, lowercase letters, numbers, and punctuation marks may be had on one template, or there are templates with only capitals and numbers. In addition to conventional sans serif letters, several typefaces are available, such as Old English, gothic, and script. There are templates with symbols used for maps, geology, mathematics, music, and other special subjects. Custom-made templates are available on special order. Numbers on the face of the template indicate the height of a vertical letter in thousandths of an inch.

To use the scriber lettering device, you must first construct a base for the template by securing a rule or T-square to the drawing board at both ends. (It is possible to use a nonskid cork-backed rule for this purpose without securing it first, but I do not recommend it for more than a minimum of lettering.) Apply two strips of masking tape to each end of the rule as shown in figure 58. This will hold the rule in position yet allow the copy to be moved freely beneath it. Each time you move the copy up, tape it down so it will not slip out of position as you letter. If the rule does not extend beyond the edges of the copy, it may be carefully taped directly to the copy. (It is easier, however, to move the copy beneath the base rule for each new line of lettering than to move and retape the taped-down rule.)

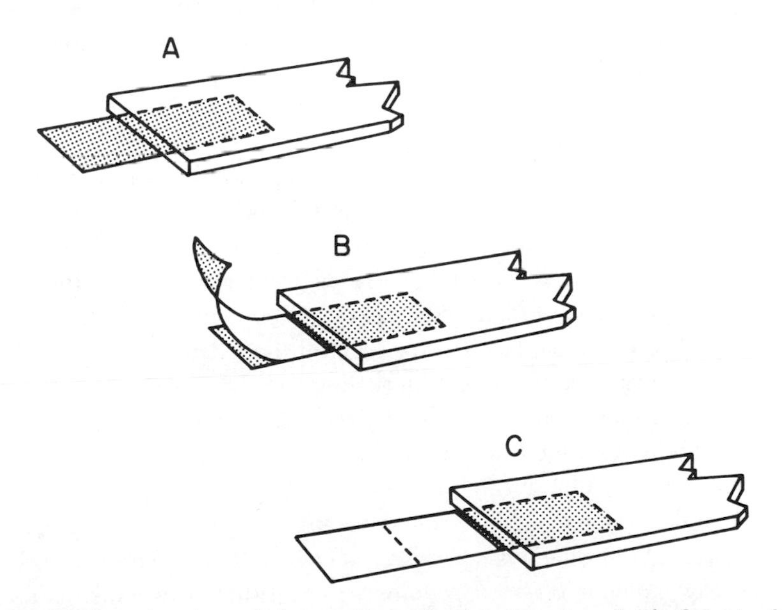

FIG. 58 Preparing a base rule for use with a scriber lettering system. A, The first strip of gummed tape is applied to the underside of the rule, with an inch extending from the end. B, The second piece of tape is applied to the end of the first piece, gummed sides together. C, The rule is taped to the drawing board, ready for paper to be slipped under it.

The size of the letters dictates the choice of pen size. The appropriate point for proper balance of line width with letter size is shown on the template.

To use a technical pen with the scriber, fit the pen into the pen arm and tighten the screw to hold it firmly. You may need to adjust the elevation screw on the scriber so the ink flows freely. If you must shake the pen slightly to start the ink flowing, do not shake it over your illustration.

A word of caution at this point: When using any lettering aid, one tends to concentrate on letters rather than on words or phrases, and spelling mistakes are likely to abound. The wise illustrator will write out the entire body of work to be lettered and follow this guide from start to finish.

With the chosen template in position against the base rule, place one point of the scriber in the desired letter of the template and the tail point in the guide groove (see fig. 11). Move the scriber so that the template point follows the depressed letter, and the letter will be drawn by the ink point. You will make fewer mistakes if you keep your eye on the point in the depressed letter instead of on the inked letter being formed. As each letter is completed, move the template along the base until the next letter is in position.

Where spacing of words and letters is critical, it may be best to letter lightly in graphite first and repeat the lettering in ink when the proper composition is achieved. This procedure may be necessary on a map, for instance, where a label is offset to the left of a point. It is possible to letter backward, but this often leads to misspelling and smears the already inked letters. Of course, if you do the lettering on Scotch 811 Magic Plus removable tape (see p. 100), all the lettering can be done conveniently and placed on the illustration later.

When a line of lettering is completed, move the copy up to the proper position for the next line. If the copy being lettered is on transparent vellum or film, placing a sheet of grid paper under it will make it easier to letter in parallel lines; otherwise use a grid-marked transparent rule or a right-angle triangle to ensure parallel lettering. Work proceeds from top to bottom on the copy to prevent smearing the wet ink.

An artist who does a large amount of varied biological illustrating will probably find the relatively expensive scriber lettering system a good investment. You can buy only the scriber and such templates as you need immediately, of course; also, other brands of templates are available.

The pens and points purchased with mechanical lettering devices must be cleaned thoroughly after each use. In general, these points are cared for the same as technical pens (see p. 70).

Preprinted Lettering

Two kinds of preprinted lettering are available on transparent sheets: press-on letters and dry-transfer letters. Both kinds come in a wide variety of type styles, in many sizes, in black and white and some colors. Numbers, symbols, stars, arrows, lines of various widths, and decorative motifs are available as well (see fig. 36). When you must letter along a curved line, preprinted lettering is easier to use than a mechanical lettering device. In general, however, it takes patience and practice to apply preprinted lettering evenly and in straight lines. You should also be aware that the potential for breaking and flaking off the copy has caused some publishers to refuse illustrations containing dry-transfer lettering. Check with your editor before you proceed.

Even in the most clumsy hands, press-on and dry-transfer lettering produce high-quality individual letters on the copy. Unfortunately, the letters often are strung together at varying angles and elevations, revealing sloppy workmanship. If you are working on a transparent ground, place it over grid paper so you can line up individual letters properly in relation to one another on both horizontal and vertical axes. If you are working on an opaque ground, rule your reference lines lightly in pale blue pencil.

Press-on letters are cut from their sheets and placed on the copy one by one. Use an X-Acto blade to cut and a small forceps to handle the cut-out letters. The transparent sheets have guidelines that aid in placing the letters evenly and spacing them correctly. Even so, you will need to use grid paper under transparent grounds and light blue guidelines on opaque grounds.

Some brands of press-on lettering are too "tacky"; the adhesive on the back adheres so strongly to the copy that it is impossible to remove a letter, or removal will take up any inked line or symbol the letter touches. Look for "low-tack" varieties when buying press-on lettering; these usually are safe, especially if all the ink is allowed to dry thoroughly before the lettering is applied. Formatt is a satisfactory brand of press-on lettering.

What is pressed on can also drop off, and press-on letters sometimes disappear from illustrations on the way to the printer. It is prudent to buy new sheets of lettering; old ones tend to dry out and in time may yellow.

Dry-transfer lettering is printed on the back of a transparent sheet. The illustrator holds the sheet of print directly in place on the copy and transfers each letter to the illustration by burnishing it down with a smooth, blunt instrument. There are special tools for this purpose, but a ballpoint pen can be used as the burnisher.

Dry-transfer lettering is notorious for flaking away, especially if the product is not new. The letters easily break apart, scattering black flecks across the copy that may not be noticed until the freckled illustration appears in print. For this reason, inspect the illustration carefully and spray it with a fixative as soon as the work is finished. As with press-on lettering, if there is any question about the age of the supply on hand, buy a fresh sheet. Chartpak is a satisfactory brand.

Preprinted lettering is not permanent. If the illustration must last longer than a year, it would be best to do the lettering with a mechanical lettering system.

8

Poster Sessions

Each year, scientific society conferences are attended by more and more researchers who wish to present papers. The time allowed at such meetings can be increased only so much to accommodate them. Consequently, poster presentations are being used more frequently to put scientific findings before colleagues.

Preliminary Planning

The first consideration is to allow plenty of time. There is no leeway; the presentation must be ready and assembled by a specific date, with no possible change in schedule. Work delays are inevitable, and time for them should be factored into planning and construction.

The illustrator must ask about the exact dimensions of the display area allowed for the poster presentation. Usually, wall space 4 feet in height and 8 feet in length is allotted to each participant, but the length may be as short as 4 feet. At some meetings the poster presentations must be displayed on easels, not hung on walls or tacked to panels.

Plan the layout of the presentation precisely. Use a table the size of the space you have been allowed, or cut a sheet of butcher paper or brown craft paper to the correct dimensions and lay this on the floor. On this dummy panel, place all the components of your poster presentation—titles, legends, photographs, tables and charts, maps, and so forth. You can also plan the presentation on graph paper to exact scale, but judging the best size of printing is more difficult with this planning method.

For convenience in transport as well as safety in handling, construct the poster presentation in sections you can carry easily. So that the sections will fit into a briefcase or other carrying case, cut the mounting boards all the same width, using the inside dimensions of the case. Some mounting boards, such as those containing legends or titles, may be narrow strips; these should be hinged to the next board in line by taping the two together on the back. The boards can be folded on the tape hinges to fit into the carrying case. At the meeting, the hinged boards can be tacked into place easily and quickly, and all the parts will be in the correct order.

Another advantage of mounting all the titles, labels, legends, and illustrations on a few boards is that there is less chance of losing parts or misplacing legends when you display the work.

If you must travel by air, all sections should fit into a briefcase or a package you can carry on the plane with you—inquire of the airline, well in advance, for size limits. (Do not trust your hard work to baggage handlers; you may well find your poster presentation winging its way to London while you stand cursing in the airport near your meeting.)

Designing the Presentation

Think of the poster presentation as an *illustrated abstract*. It should contain the maximum information with the minimum number of words. This is where illustrations will carry your story. As an example, look at a good museum exhibit: the pictures (model) show the information; explanatory legends are short, to the point, and legible at a comfortable distance.

The poster presentation not only must catch the attention of the passerby but must maintain interest until the information has been read. Making the display easy on the viewer helps accomplish this. People tend to stop in groups, and there will be the inevitable viewer (a possible employer) who has forgotten to bring eyeglasses; these people should be able to see and read the material without strain.

Graphics such as maps, graphs, and tables should be clearly visible, easily read, and as brief as possible. Information should be given in phrases and ideas, not in run-on paragraphs, and must be legible at a distance of 4 or 5 feet. As a help in deciding what size lettering is best, cut out a page of advertising, tape it to the wall, and step back 4 or 5 feet to look at it. The smallest lettering you can see well from that distance is the smallest you should use on your poster.

Photographs and drawings should also be large enough to be seen clearly at this distance. Matte-finish photographs are less reflective and more attractive for display. If you use enlargements from photocopy machines, make the copies on heavier paper than usual.

Drawings should have enough contrast, with bold enough lines and shading, to be seen clearly without squinting. If there are labels on the drawings, these must also be legible from 4 to 5 feet away.

Lettering

Labels and legends may be lettered directly on the mat boards on which the illustrations are mounted, or they may be done on separate strips of paper or board and either fixed to the mounting boards or positioned above or below them.

SIZE OF LETTERING

The most important information, the title of the presentation, should be 1½ to 2 inches high, all in capital letters. The abstract number also is this height and usually is placed in the upper left corner.

Affiliations are lettered slightly smaller, about one inch high. These include the author(s), university or other institution, and addresses.

The rest of the lettering and numbering should be readable from 4 to 5 feet away. Headings will be slightly larger or bolder than the type used for the body of information.

Remember that it is easier to read short lines than longer, run-on text. At normal book-reading distance (about 13 inches), a line of print should be no longer than 5 inches. Multiply this by the distance from your expected viewers (about 4 feet), and you will find that no line on your poster should run longer than about 20 inches. Crisp phrases arranged in columns may be best; on posters it is permissible to omit less important words such as *the* or *and* ("telegraphic" style).

METHODS OF LETTERING

The budget may influence the method of lettering used. The method chosen may also depend on other factors, such as whether the illustrator is seeking a job at the meeting; in this case handsome but costly lettering may be worth the extra expense.

The most attractive lettering is typeset to order. Many type styles are available, done in any size. Avoid decorative styles with curling serifs and peculiar angles; these are hard to read from any distance. The cost of typeset letters is considerable.

Dry-transfer and adhesive vinyl letters come in large sizes, useful for titles and abstract numbers, but these sizes may have to be ordered in advance. Some colors as well as black and white are available. Choose a simple style, such as Helvetica, to harmonize with lettering in the rest of the presentation.

In all dry-transfer lettering, the letters are fragile. If the supply is not fresh, the letters will crack and flake off the board; return the package to the store for replacement (one of the inevitable delays mentioned above). When the dry-transfer lettering has been completed to your satisfaction, spray the work with fixative to protect it from accidental scrapes.

If dry-transfer lettering is done directly on the mat board on which illustrations are mounted, it will require great care to apply it in a straight line, and any drawn guidelines may be evident on the finished work. An alternative is to do the lettering on two-ply plate-finish paper with a light (tracing) table; place a sheet of grid paper under the plate-finish paper for guidelines. Cut out the lettered word or phrase and glue it in place on the mat board. In this case it would be better to use a medium-colored background, since white plate-finish paper shows up glaringly against dark or vivid mat boards.

Large lettering can be done by stencil. Several styles are available, but it is best to choose a simple, easy-to-read type. When using a stencil, you can letter in any color. For legibility, use a darker color and fill in the outlines. For lettering that demands attention, you could stencil the title on an intense-colored art paper, cut out the letters, and fix these to the presentation mat board.

Scriber lettering systems (such as the Leroy system) can be used to make title-sized letters. There are large-sized templates, for a large price. An alternative method is to use the largest size available to you, making sure the letters are densely black and clean edged, then have this lettering enlarged by photocopy to the proper size.

You can letter by scriber system directly on the mat board used to mount the photograph or other illustration, but this can lead to frustration if a word is found misspelled or missing. It may be easier to letter the legend or label on a separate piece of paper to be fixed to the mat board. Or letter on tracing vellum over grid paper, then photocopy the tracing, cut out the lettering, and fix it to the mat board.

Scriber lettering may be done conveniently in any length line on Scotch 811 Magic Plus removable tape (see p. 100 for method and precautions). The tape then is cut where desired and placed in position on the mounting boards. The tape has a slight sheen that may be apparent at certain angles.

With the Kroy lettering system, available in some university science departments, a wide variety of lettering sizes and styles can be produced with ease. Directions come with the device. Again, be aware of a possible sheen.

Another technique for labeling the poster presentation is computer-generated lettering. It may be well worth your time to locate and learn to use whatever system is available to you. Be sure the lettering is dark enough to be clearly legible from 4 to 5 feet away.

Mounting the Presentation

Photographs and other illustrations may be mounted on boards by one of several methods, as discussed in chapter 10. Use a heavier mounting board if the poster is to be hung or displayed on an easel; use a lightweight board if it must be tacked to a backboard. An excellent mounting board with a layer of foam sandwiched in the middle can be bought at most art-supply stores. This is lightweight, relatively sturdy, and comes in various thicknesses and sometimes in a choice of colors.

The illustrator should take along a roll of butcher paper (any unobtrusive color) to cover the display area. Too often the tack boards provided are covered with holes and scribbling, making an amateurish background for your difficult and brilliant work. Take also a supply of heavy tacks, twine, invisible mending tape, and strong adhesive tape. Scissors, an X-Acto knife, several technical pens with black ink, and a bottle of white-out paint and a brush are also useful to have on hand.

9

Photographs, Plates, and Projection Slides

This chapter covers retouching photographs, lettering on photographs, and their proper arrangement on a plate. It does not treat the use of photographs, as such, for illustrations. In addition, I discuss problems peculiar to producing graphics to be photographed for projection slides.

Photographs

Handle photographs as little as possible and avoid fingerprints, which tend to attract dust and dirt and show up in reproduction. Paper clips may destroy the emulsion on the front (if you must use paper clips, sandwich the photograph between two index cards first). Writing on the back of a photograph may cause ridges on the front that are obvious in reproduction; use soft pencil or grease pencil if you must write on the back, and place the photograph on a hard surface while writing. Alternatively, type a label and tape this to the back of the photograph with removable tape.

RETOUCHING

Minor retouching is not difficult, but unless you are skilled in the use of an airbrush, do not attempt to cover large areas. Very detailed photographs usually are not reduced for publication, but in retouching it is better to work on a print twice the expected final size. This amount of reduction (one-half off) will minimize any obvious brush or pencil strokes. Although they do not reproduce as sharply as glossy photographs, matte-surface prints are easier to retouch.

You will need at least two prints; one will be the working print, and the second will serve as a reference showing the progress of the ongoing work. Additional prints may come in handy as spares in the event of an unsatisfactory first rendition.

Special retouching kits are sold at large photographic-supply stores. Such kits contain dyes appropriate to different photographic-paper emulsions and instructions giving exact details on using the dyes. Retouch gray paints are available in warm, cool, and neutral tints, in a wide value range, such as those made by Grumbacher and Pelikan. These are similar to gouache paints in their consistency and water-solubility, but they contain an anticrawl agent that prevents their beading up on the photograph. Such retouch products may be used on glossy as well as matte prints. In addition to paints, you will need eyedroppers, mixing dishes, and good-quality watercolor brushes.

Photographs may be printed in a slightly blue or brown tint; match the hue carefully, since unmatched paint is glaringly obvious. Dyes and paints should be applied with a fine brush, stippling carefully to match values. Large areas of flat tone will be unattractive in the printed result. Small areas can be blotted and lightly smudged with absorbent cotton or cleansing tissue to blend the edges of the paint.

If retouch dyes or paints are not available, you may use other materials and methods. On matte-surface prints, use hard- and soft-lead pencils, charcoal pencils, and white Conte pencils to emphasize or minimize textures and highlights and to bring obscure parts of the photograph into prominence (see fig. 59A,B). A kneaded eraser is the least damaging kind.

Hold the pencil lightly, and touch the photograph with soft, short brushing strokes. To blend the individual strokes, hold the pencil at a low angle and use a gentle, sweeping motion. To darken this area, apply another layer of graphite softly over the first. (To lighten a dark area on the photograph, use the white Conte pencil in this same manner.) This gradual building up lessens the chance of mistakes and erasures, which tend to make an illustration look ragged. Careful smudging—with a rolled-paper smudger, not your finger—may be done sparingly; too much can give a dull and amateurish appearance. A light application of spray fixative will protect the work from dust and accidental smudging.

Pencils usually give better results, but matte-surface photographs may also be retouched with gouache paints. Match the cool or warm

FIG. 59A Unretouched photograph of a fossil frog, *Eopelobates grandis*. (Courtesy American Museum of Natural History)

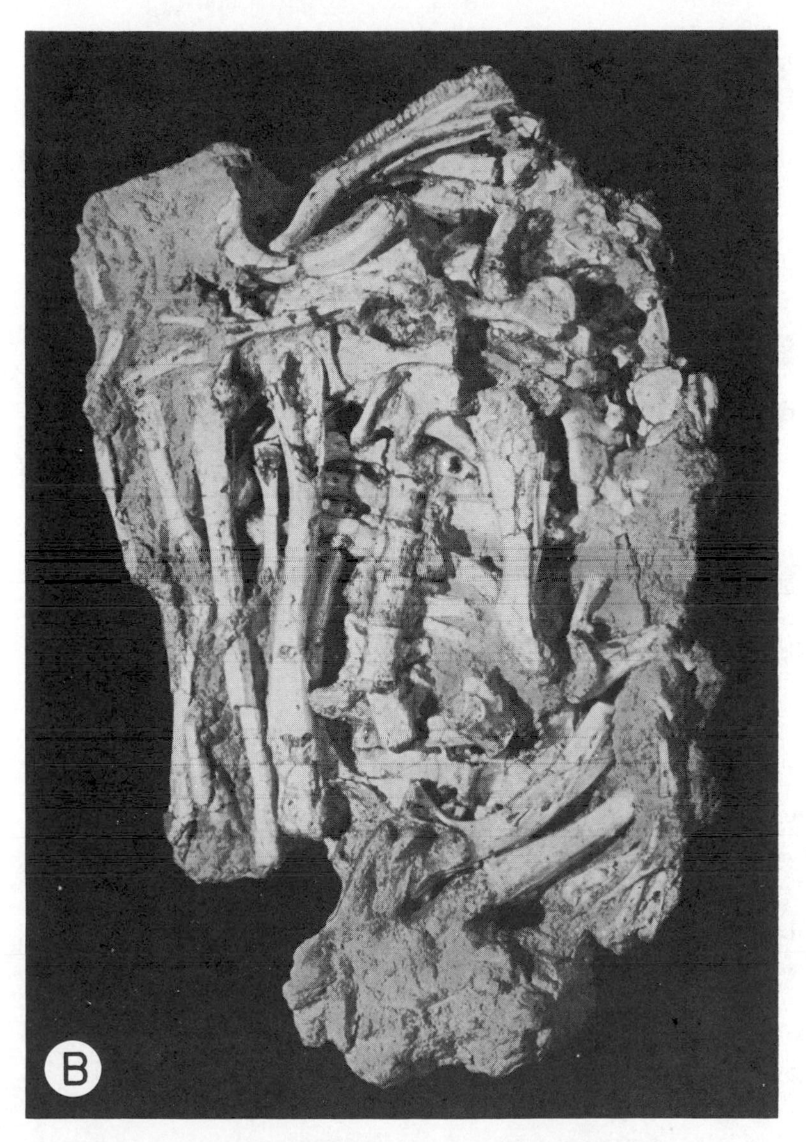

FIG. 59B The same photograph as figure 59A, retouched to help differentiate between bone and matrix. (Courtesy American Museum of Natural History)

(blue or brown) tint of the photograph, and use a fine brush to stipple the paint. Again, small areas may be blotted and lightly smudged to blend the edges of the paint. To remove either gouache paints or retouch paints, wet and blot—do not scrub the surface of the photograph. Wait for the surface to dry before attempting to retouch again.

Waterproof india ink may be used to draw lines and dots on matte prints. Keep in mind, however, that all lines will be broken into a series of dots by the screen used in halftone reproduction.

Glossy prints are decidedly more difficult to retouch than matte prints, and you must work with retouch or gouache paints, not with pencils. If you use gouache paints, the gloss on the photograph can be diminished somewhat by rubbing the surface with special dulling powder, called "pounce," or with a dulling (matte) spray film.

Apply the gouache paint carefully and sparingly; it should not be too thin. If possible, use an extra print to try out strokes and blending techniques before you work on the final copy. Blotting the wet paint with an absorbent tissue will lessen the "painty" appearance, or you can blend it into the picture with your fingertip. The paint may be "erased" by washing it out with a brush and clear water, but blot the wet paper immediately to prevent warping.

Special inks are available for drawing lines and dots on glossy surfaces. In an emergency, you may be able to use india ink mixed with soap.

SILHOUETTES

It may be desirable to remove background material in a photograph so the subject stands out clearly against white or black. This can be done by painting over the entire background of the photograph, but the brush strokes will show in the halftone reproduction. It is usually better to have silhouetting done by the printer, even though this adds to the cost.

The illustrator must indicate to the printer the edges of the figure to be silhouetted. Paint a band of opaque white or black around the periphery of the subject, taking care to follow the margin exactly. The paint must be dense enough that no trace of the underlying photograph shows through. The figure is then delivered to the publisher with instructions to prepare the plate as a silhouette. Using the painted margin as a guide, the printer will remove the background material (fig. 60A,B).

FIG. 60 Silhouetting a photograph to remove unwanted background. A, One-half of the lizard (*Cnemidophorus burti stictogrammus*) has been carefully outlined with superwhite gouache paint. B, After the lizard was completely outlined, the printer was able to remove the rest of the background. (Courtesy American Museum of Natural History)

An alternative method of silhouetting is to make an overlay showing the area to be eliminated (see p. 96 for making an overlay). Place register marks on the illustration outside the image area and on the clear polyester overlay sheet. Mark the subject area precisely, using ink made for use with polyester. Fill in this area with ink. The printer will drop out everything except the part you have covered in black. Masking films are available at art-supply stores, coated in transparent red or amber. The area to be silhouetted is designated by cutting through the thin film of color with an X-Acto knife and peeling away the unwanted color. This method of preparing a silhouette will somewhat reduce the printer's work and the cost.

LETTERING

Numbering and lettering on photographs can be done in several ways. As long as the work is neat and clean, the result will be acceptable to most editors.

It is possible to letter directly on the photograph, using an ink made especially for glossy surfaces. Black ink will disappear against a black background; you may need to highlight the edges of the numbers with white or use white ink in dark areas.

Letters and numbers may be drawn on a separate piece of paper and then cut out and glued into place. The letters will appear in the center of a clear white area (the "white" will actually have a faint dot pattern, since it will be screened with the rest of the halftone). The edges of the glued slip of paper may show as a pale shadow in reproduction. If the work is done neatly, with the edges of the slip truly square and all traces of excess glue cleaned away, there would be no reason to reject such a plate for its appearance.

There are dangers, however, that should be considered. Small slips of paper can easily be rubbed off and lost. They can twist out of position on the plate, making the illustration look sloppy and amateurish. In addition, most glues will dry up in a year (or less) and may turn yellow and stain the photograph. Using double-sided adhesive tape in place of glue generally produces a more permanent and neater result. Apply the printed labels, numbers, and such, to the tape and stick this to a slip of waxed paper or clear acetate to determine the correct position on the copy. Carefully transfer the taped-down label to the copy, using forceps for ease and cleanliness.

Lettering can also be done on transparent film, cut out, and placed on the photograph. Scotch 811 Magic Plus removable tape is one such product (see p. 100), as is the Kroy lettering device tape (see p. 101). In spite of the fineness of these films, the edges may show as faint shadows in halftone reproduction. The films may have a little gloss, too, which may appear as a slight sheen (no edges or sheen will appear in line-copy reproduction).

Commercial press-on lettering (see p. 105) is convenient. There are many sizes and styles, including letters, numbers, and various symbols. These are available in black or white, some with shaded or highlighted edges, some inside circles to make them stand out from the background. There are also lines and arrows, shaded or plain. The film on which these all are printed, though extremely thin, may show a faint shadowed edge in halftone reproductions. Only fresh products should be used, since old stock tends to lose its adhesive qualities and may pop off before the copy reaches the printer.

Dry-transfer lettering (see p. 106), also convenient and attractive, shows no shadowed edges in halftone reproductions. It has other problems, however. The print easily breaks and flakes away and must be sprayed with a fixative for safety. Such a spray may alter the appearance of the photograph; try it on a spare copy before you submit your good illustration to this treatment. All mounted photographs should be protected from grease and dust with a paper cover or flap (see p. 128); but such covers may rub against photographs and pick up dry-transfer lettering in spite of a protective spray.

For the most professional look, all lettering can be done on an overlay (see p. 96). The printer is told to drop it out from the image (so it will show as white print on the photograph) or to print it in black, as desired. This adds to the cost of reproduction, of course.

Each method of lettering on photographs has its advantages and its difficulties. The illustrator's choice may ultimately be decided by availability and budget.

Arranging Figures into Plates

Often it is necessary to combine several photographs (or other tone illustrations) or individual line drawings into a single plate, as when two or more illustrations are best viewed at once. In the case of photographs, money can be saved if photographs are grouped rather than presented individually on separate pages of a publication. One can give the printer several loose figures—photographs or drawings—together with a diagram of how they are to be arranged and instructions for adjusting the relative sizes of individual figures if needed (a "dummy"; see p. 13). However, it is better and less expensive to provide camera-ready plates.

As a general rule, photographs to be mounted side by side on a plate should have the same vertical dimensions. A plate looks best if all components are the same size, though one can vary the horizontal dimensions of two or more side-by-side figures so long as their total width conforms to the common width established for the other figures on the plate. Group the photographs closely; there is no point in paying to publish large white areas between them. If you have a good eye and your photographs are truly rectangular (use a razor to cut photographs and to trim away the white borders), you can mount them leaving an even, narrow (ca. 1 mm) white space between adjacent figures. Since there may be a slight shadow line in the space, most editors prefer that you abut figures and then direct the printer (at extra cost) to engrave a precise white space between them.

Combining line drawings (black-and-white illustrations) into plates differs from assembling photographs in that there is usually no background to consider and so no edges to align or abut. (If there are clearly horizontal or vertical aspects to the figures, be certain they are properly aligned.) Before reaching the stage of assembling a plate, the illustrator should have anticipated this and made the drawings with a common reduction in mind; the separate figures should be drawn in the same style and with similar line widths to give the plate an appearance of unity. Circumstances other than appearance may dictate the arrangement of components of a plate, but if this is not so, try to achieve a balanced arrangement (for example, put the denser, heavier figures at the bottom). Plates are a place where the lighting convention is important: figures will look better and be easier to compare if they all appear to be lighted from a common direction.

Where there are only two elements to a plate, the caption may simply refer to left and right or upper and lower. More often it will be necessary to distinguish the various parts with letters placed in a standard position on each photograph or beside each line figure. Alternatively, you can mount photographs with extra space between them vertically and letter on that white background. This may provide better definition of the letters, but the waste of space and the cost of adding another step in the printing (a linecut will be needed besides the halftone) weigh against this method (see p. 5).

Slides

Just as with a published report, the quality of the graphics is an important factor in the slide-show audience's understanding of the work being presented. And justified or not, it can influence a viewer's perception of the presenter's respect for his audience and the quality of his research. It is important to appreciate that a figure or table perfectly adequate for publication in print may be a disaster as a projection slide. If anything, materials designed for slides require more thought than those to be printed. A reader can take time to puzzle over an unclear diagram and even resort to a magnifying glass if the details are too small. Viewers of slides may have only seconds to take in what is on the screen, and if you make it too difficult for them, the soporific influence of a darkened room and a gently whirring projector can lose you your audience.

This section addresses the preparation of copy to be photographed for projection slides; photography itself is not treated. A variety of sizes and shapes of slides have been used in the past, but the virtually universal standard now is a square cardboard frame measuring 2 inches on a side, holding a photographic transparency made on 35 millimeter film. The conditions under which a slide is projected can vary greatly depending on such factors as the adequacy of the projector and screen for the size of the room. Unless specifically stated otherwise, the discussion is directed at copy for this sort of slide projected under at least minimally adequate conditions.

NATURE OF COPY

Graphic projection slides can range from solid blocks of text to elaborate drawings. Whatever the material, the author of the text and the illustrator must keep simplicity in mind. Write in telegraphic style rather than in complete sentences; do not force more information into a figure than is necessary and can be comprehended in a brief viewing; use several slides rather than one if this will help.

COPY ORIENTATION

Though the slide itself is square, the projected photograph is a rectangle with the long axis 1.5 times the short axis (33 × 22 mm on the film). Thus the copy can be oriented either way. If possible, the copy should be drafted with the long axis horizontal, mainly because projectionists often set up to fill the screen with an image oriented horizontally. When the occasional vertical slide is projected, the top and bottom edges may be truncated. A second reason is that a vertical slide properly projected will usually have a smaller image than a horizontal one could have on the same screen, since screens commonly are wider than they are high.

SIZE OF COPY

You can work at any convenient size, since the size of the artwork (within reason) is of no great consequence to the photographer. As with other artwork, however, it is convenient to stay within the limits of the standard letter paper size of 8.5 by 11 inches. Allowing for mounting on board of this size, you might aim for copy 10 inches wide. Applying the 1.5 to 1 ratio of width to height, this would allow you a vertical dimension for the copy of about $6\frac{5}{8}$ to $6\frac{3}{4}$ inches if the space is to be used fully. The chief importance of size of the copy is that this dictates the minimum size of lettering you should use.

LETTERING

In lettering on copy for a slide, there are several points to consider:

1. Use a simple style, without serifs or fancy curls, for best legibility. The Leroy style is suitable.

2. Capital letters are easier to read when projected on a screen. In addition, they tend to close up less than round lowercase letters.

3. Do all lettering horizontally, if possible, or at least in only two directions. It is asking a lot of an audience to require craning necks at several angles.

4. The smallest letter (or number) on the slide will determine the size of the rest. This smallest letter size depends on the size and shape of the original copy (see below).

5. Use a pen size (width of line) appropriate for the size of lettering chosen. If you letter with the Leroy system, use the pen size recommended for each template. Fine lines usually do not project well, becoming blurred or completely lost. Bold print and lines are preferred.

To be legible to the audience when projected, the lettering on a slide must be at least a certain size. Hence the original artwork must adhere to the same ratio. The following table gives the minimum permissible letter size for a range of widths of copy. Remember to measure the *smallest* letters—do not measure capitals and then use lowercase or neglect superscripts or subscripts that may project as tiny spots.

Copy Width in Inches	*Letter Height in Millimeters*
4	2.0
6	2.4
8	3.2
10	4.0
12	5.0
14	5.7
16	6.5
18	7.2
20	8.0
22	8.8

If the slide is designed to be square or vertical rather than horizontal, measure the height in inches, as above, but increase the letter size slightly. For example, copy measuring 6 inches high may require lettering of at least 3 millimeters; vertical copy 18 inches high will take lettering of no less than 8 millimeters.

A typewriter with a fabric ribbon produces copy that photographs and projects poorly. Typing with a one-time-use carbon ribbon can, with respect for copy-width limitations imposed by the type size, produce adequate copy. Pica size type (ten characters to the inch), for example, measures about 3 millimeters high and is satisfactory for copy up to about 7.5 inches wide. If several type styles are available, choose the heaviest.

A border generally should not be used around material presented on a slide. The frame of the slide will be border enough, and a printed border will use up valuable space.

You can make your slides more attractive and attention getting by using color. Dry-transfer lettering comes in a variety of colors and may be used tastefully for titles and headings. Or use a colored background with black printing (be cautious about using too dark a background, or the print may be unreadable). When placing one color against another, choose colors with a distinct difference in value, such as yellow and dark blue. If the background and lettering colors are too close in value, the letters may seem to shimmer or be fuzzy.

10

Mounting and Packing Illustrations

Finished illustrations must be protected from damage or soiling. This is not difficult or expensive, and it is as much a part of the illustrator's job as making the illustration. In addition, illustrations need to be documented and, if shipped, properly packed.

Mounting Illustrations

An illustration done on a fragile surface or one that is likely to warp should be attached to a mounting board such as stiff white poster board, illustration board, or foam-core mounting board. This board must be cut somewhat larger than the illustration itself. For ease in handling and to prevent loss of small copy, all the illustrations for one project should be mounted on the same size of mounting board. Determine the largest board needed, and cut as many pieces of this size as there are separate illustrations.

If all the illustrations submitted with a manuscript are smaller than a manuscript page, cut the mounting boards to 8½ by 11 inches (the size of the manuscript paper), and pack illustrations and manuscript together.

The choice of adhesive depends primarily on whether permanent or temporary mounting is required. For temporary attachment rubber cement, applied with brush or pressure can, is most widely used. It is definitely temporary and should not be used when the mounting will last more than a year. In addition to becoming dry, it will eventually turn yellow and may stain the illustration. For temporary mounting, however, rubber cement is quite satisfactory.

Place the illustration on the chosen backing board and trace around it very lightly. Apply a thin coat of rubber cement to the back of the illustration and also to the area marked on the mounting board. Allow the adhesive coats to dry for a few minutes, then press the illustration gently and smoothly into place. Allow this to set a short while, then, with light pressure and a clean cloth, rub off any excess cement. (A dry blob of rubber cement also makes an excellent remover.) Since rubber cement will yellow and stain with age, be sure to remove every trace from the illustration surface.

An illustration mounted in this manner will lie flat for reproduction but can be removed from the mount if necessary. To remove copy, moisten a corner of the illustration with rubber-cement thinner and carefully peel back a tiny bit. Gently peel this away from the mounting board, applying thinner as you go. (Use caution; the thinner is highly flammable.) Allow the illustration to dry a little, then wipe off any remaining cement. Flatten the illustration under light pressure if necessary.

Double-sided adhesive film is an excellent medium for smooth, strong mounting of illustrations. The thin sheets are protected on both sides by paper, which is peeled away to expose the adhesive surface. For best results and to reduce frustration, apply double-sided adhesive film in the following manner.

Fixing the Adhesive Sheet to the Illustration

1. Cut the sheet slightly larger than the copy to be mounted.
2. Partly peel the protective paper back from one end of one side of the adhesive film.
3. Carefully align the upper edge of the back side of the copy on the exposed adhesive without any overlap; press the copy down smoothly.
4. Slowly peel away the rest of the protective backing, smoothing the copy into place and pressing out any air bubbles as you go (the second side of the adhesive film will still be covered with protective backing at this point).
5. With a razor, trim the adhesive sheet to fit the copy exactly.

Fixing the Illustration to the Mounting Board

1. Place the copy on the board and lightly mark its correct position.
2. Turn the copy face down and peel back about an inch of the remaining protective sheet.

3. Turn the copy right side up and position it on the marks made previously on the mounting board; smooth the exposed inch of adhesive to the mounting board.
4. Progressively peel away the backing while pressing the copy to the mounting board, smoothing away any air bubbles as you go.

It is important to remove the protective backing gradually, not all at once, lest the exposed adhesive sheet flip over upon itself and become hopelessly stuck.

Permanent and smooth mounting can also be achieved with dry-mounting tissue, available in sheets and rolls at photographic- and art-supply stores. Dry mounting requires a heat source. There are special presses for this, but a household iron can give excellent results.

Mark the position of the illustration on the mounting board. Cut a sheet of tissue exactly to fit the copy to be mounted. Tack the tissue to the back of the copy at two corners with the warm iron. Place the illustration, with attached tissue, in position on the mounting board. Tack down the same two corners, with a protective slip of paper placed between the iron and the face of the illustration. Then place the mounted copy face up on a blanket or layer of felt on a flat surface. With a clean blotter covering the illustration, press the iron, heated to 175°F, down on the blotter for one minute. Weight the mounted illustration until it cools.

This method of mounting is equally well adapted for use with photographs and several artist's grounds. It should not be used with grease pencil (wax crayon) drawings, such as those done on coquille board or stipple-surface scratchboard. Wax-back shading films and dry-transfer lettering also will be ruined by the dry-mounting procedure. There are special dry-mounting tissues for use with acetate films, however; directions come with the purchase. Photographs can be mounted successfully with the dry-mounting process, but excessive heat may destroy the emulsion. Follow the directions provided with the tissue for best results. Try out a test piece before you subject your illustration to this method.

It is not always necessary to fix the entire area of the illustration to the mounting board. Sometimes just taping the edges securely is enough. Graphs and maps may be taped at top and bottom, using white artist's tape or Scotch Magic tape. If the printer specifies a flexible ground, to be wrapped on a drum for printing, tape the illustration only at the top. Since grounds and mounting boards may expand or shrink at different rates as humidity changes, taping the illustration only at the top will prevent buckling.

Protecting Illustrations

Pencil drawings are easily smudged; a careless scratch may disfigure a scratchboard drawing; accumulated dust, dirt, and spilled ink can destroy the work of hours or days. It is essential that an illustration, once completed, be protected against these dangers.

A widely used form of protection for pencil, ink, and carbon-dust drawings is fixative in aerosol cans. The fixative is sprayed from a distance of 1 to 2 feet onto the upright illustration, sweeping lightly over the surface with several applications. Always test the spray on the same kind of ground treated with the same art technique as the illustration before trusting your artwork to it. Some sprays will cause pencil to run or spot or make carbon dust fuzz or puddle. The aerosol can may be faulty, spritzing out blobs of liquid instead of a fine mist. It takes only seconds to test, and you may save hours of redrawing.

In addition to spraying on fixative, you must provide cover sheets ("flaps") as protection for illustrations. Heavy brown wrapping paper, strong white bond, and acetate film are all suitable. Cut the cover paper to the width of the mounted illustration and 2 to 3 inches more than its height. With white artist's tape or Scotch Magic tape, attach the paper to the back of the mounting board near the top and fold it over the front of the illustration. If the illustration is to be referred to frequently, the cover paper may be trimmed flush with the bottom edge of the board and left free; otherwise it is folded under and taped in place. An acetate cover is best cut to the exact size of the mounting board and taped down at top and bottom.

Frisket is a clear film with a low-tack adhesive on one side. It can be used to protect photographs and inked illustrations from dust and fingerprints, but it will pick up pencil and carbon dust. Test the frisket on a scrap before applying it to an illustration; if the tack is not "low" enough, it could damage the work (dry-transfer lettering and symbols are especially vulnerable). "Removable" frisket is safest.

Documenting Illustrations

Every illustration should have an identifying phrase or legend written on it or, when mounted, on the board. This is particularly important where confusion among specimens illustrated is possible.

You may write in lead pencil on the back of an illustration, but any writing on the face of the drawing or mounting board must be done in the margin with light blue pencil. An exception is marking the top of the illustration. "Top" should be written in black, outside the image area, so it will appear on the negative to aid the printer; the word will be stripped out later.

Avoid writing on the back of a photograph, if possible, and using paper clips or staples (see cautions, p. 112).

Identification on an illustration should include the names of the artist and the person for whom the work was done, the publication the illustration is intended for, the figure or plate number, and the intended reduction for publication. Printers may be unable to tell top from bottom in a biological illustration; you should indicate the top of the illustration on the face of the copy in black (see above).

Packing and Mailing

Important—be sure to photocopy all illustrations before you send any originals. You will want a record of your work; and in case of loss or damage, you will have exact copies from which to reconstruct the originals.

When sending illustrations by mail, assume that "Fragile" and "Handle with Care" labels will be ignored and pack accordingly. Illustrations mailed in a large manilla envelope should be protected back and front with stiff cardboard or posterboard cut to fit snugly inside the envelope.

Before packing a larger group of illustrations, be aware that both the United States Postal Service and United Parcel Service limit the dimensions and weight of a package: length and girth added together (fig. 61) must not equal more than 108 inches; weight must not exceed 70 pounds.

A large package of illustrations should be packed in a strong box if possible. Wrap the inner package of artwork, especially illustrations done on fragile grounds such as scratchboard and coquille board, in plastic bubble film or other padding. Your package probably will not encounter water; nevertheless, it would be prudent to enclose the inner package of illustrations in a plastic bag and seal this as tightly as possible. Before sealing the outer box, place inside it a card with the sender's and receiver's names and addresses, or print this information on the inside of the box lid.

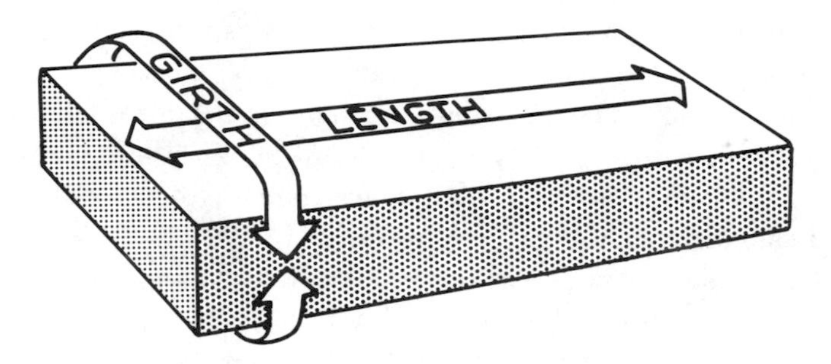

FIG. 61 Measuring the length and girth of a package.

If no proper box is available, most work done on standard grounds will survive mailing when sandwiched between layers of heavy corrugated cardboard cut at least 2 inches larger all around than the mounted illustrations. Pad fragile grounds inside the sandwich.

United Parcel Service does not allow the use of outside wrapping paper on a package (if the paper should tear, the address will be lost). Tape all around the edges of a "sandwich" with reinforced tape; tape a box securely. Do not tie the package with string, since this tends to catch and tear en route. Use permanent felt marker or waterproof ink to print the address. The final step is to insure and certify your package, and you are finished.

Afterword

The achievement of accuracy together with artistry in a drawing is one of the most challenging and satisfying aspects of scientific illustrating. It would not be surprising if a beginning illustrator were to find this challenge so enjoyable that the primary interest shifted from biology to biological illustration—it has happened before.

If you are interested in pursuing this subject further, or even considering it as a possible vocation, you may write to the following organization for information and brochures:

The Guild of Natural Science Illustrators
Post Office Box 652
Ben Franklin Station
Washington, D.C. 20044

I wish to express appreciation to the GNSI for their help over the past years, both in their informative and useful newsletters and in their general attitude of willing helpfulness and comradely support.

Selected References

Books

Allen, Arley. 1986. *Steps toward better scientific illustrations*. Lawrence, Kans.: Allen Press.

Bruno, Michael H., ed. 1986. *Pocket pal: A graphic arts production handbook*. New York: International Paper Company.

Downey, John C., and James L. Kelly. 1977. *Techniques and exercises with the pen in biological illustration*. Cedar Falls: University of Northern Iowa.

Henning, Fritz. 1986. *Drawing and painting with ink*. Cincinnati: North Light Books.

Holmes, Nigel. 1984. *Designer's guide to creating charts and diagrams*. New York: Watson-Guptill.

Holmgren, Noel H., and Bobbi Angell. 1986. *Botanical illustration: Preparation for publication*. Bronx, N.Y.: New York Botanical Garden.

Jastrzebski, Zbigniew T. 1985. *Scientific illustration: A guide for the beginning artist*. Englewood Cliffs, N.J.: Prentice-Hall.

Leslie, Clare Walker. 1980. *Nature drawing: A tool for learning*. Englewood Cliffs, N.J.: Prentice-Hall.

McCann, Michael. 1975. *Health hazards for artists*. New York: Foundation for the Community of Artists.

———. 1979. *Artist beware: The hazards and precautions in working with art and craft materials*. New York: Watson-Guptill.

Mayer, Ralph. 1979. *The artist's handbook of materials and techniques*. New York: Viking.

Montague, John. 1985. *Basic perspective drawing: A visual approach*. New York: Van Nostrand Reinhold.

Papp, Charles. 1976. *Manual of scientific illustration*. Sacramento: American Visual Aid Books.

Stone, Bernard, and Arthur Eckstein. 1983. *Preparing art for printing*. New York: Van Nostrand Reinhold.

Wood, Phyllis. 1982. *Scientific illustration: A guide to biological, zoological, and medical rendering techniques, design, printing, and display.* New York: Van Nostrand Reinhold.

Art and Equipment Catalogs

A. I. Friedman
25 West 45th Street
New York, New York 10036

Ben Meadows Company
General Catalogue
P.O. Box 80549
Atlanta, Georgia 30366

Dick Blick
14339 Michigan Avenue
Dearborn, Michigan 48120

Lee's Art Shop
220 West 57th Street
New York, New York 10019

Sam Flax
1699 Market Street
San Francisco, California 94103

Visual Systems, Inc.
1727 I Street, NW
Washington, D.C. 20006

Source for EssDee British Scraperboard

Manufactured by
British Process Boards, Ltd.
17 Lisle Avenue
Kidderminster
Worcestershire DY11 7DE
England

In the United States, may be obtained through any art-supply dealer who carries Morilla, Inc., art products.

Index

Page numbers for illustrations are in boldface